CLINICAL TRIAL

An ALS Memoir of Science, Hope, and Love

Robert K. Ranum

Print ISBN: 978-1-64719-890-9
Ebook ISBN: 978-1-64719-891-6

Published by BookLocker.com, Inc., Trenton, Georgia.

Printed on acid-free paper.

BookLocker.com, Inc.
2021

First Edition

Library of Congress Cataloguing in Publication Data
Ranum, Robert K.
CLINICAL TRIAL: An ALS Memoir of Science, Hope, and Love by Robert K. Ranum
Library of Congress Control Number: 2021921814

Ah, but a man's reach should exceed his grasp,

Or what's a heaven for?

from "Andrea del Sarto" by Robert Browning

Contents

Prologue

For an honest person, keeping a secret is like walking with a pebble in your shoe. Depending on its size and location underfoot, it can be either an annoyance or a problem of such mind-consuming proportions that it becomes a disability that forces the person to sit down, take off the shoe, and address the problem. In 2016, Dr. Laura Ranum, Director of the Center for NeuroGenetics at the University of Florida, had a secret that was becoming a disability. The man she loved most in the world and the father of her two children had been diagnosed with amyotrophic lateral sclerosis (ALS), a devastating and fatal neurological disease. That's me. I am Laura's husband. I had been diagnosed with ALS in April 2016, and although Laura's work focused on neurological diseases including ALS, she hadn't told most of the graduate students, post-docs, and other professors in the Center about my diagnosis.

The secret was gnawing away at her. The fact that her husband had ALS had become the motivating force for her work, her creative abilities, and, in many ways, her identity, yet she hadn't told anyone beyond a few close friends. Every interaction at work left her feeling as though she had failed to disclose the issue most relevant to what they were working on.

The diagnosis was just the most recent of many steps that had become progressively more worrisome for Laura. My mother had died of ALS in 1989, the third of her siblings to succumb to the disease. When that happened, Laura was hoping the gene for the disease was recessive, meaning that even if I inherited it, the normal gene from my father would protect me from developing the disease. If, however, the gene were dominant, then it would require only one copy of the gene to cause the disease, and Laura knew that I would have a 50% chance of having inherited that gene and developing ALS. Laura didn't know what the gene was, but her genetics training told her that as long as all of the affected individuals were in one generation, it could be a recessive gene.

That bubble burst with a phone call in about 1995. Laura was in the bedroom when my cousin Marilyn called. Marilyn's mother had also died from ALS, in 1976. Laura picked up the landline phone on the bedside table and, after the pleasantries, Marilyn said, "Margorie's been diagnosed with ALS." Margorie was Marilyn's sister. Laura knew immediately that meant the gene was dominant and therefore threatened me and our two kids. A wave of nausea swept over her, and after the call, she ran to the bathroom and threw up.

Because Laura worked with neurological diseases, she knew what damage they could do to a family. Worry simmered in her for years before reaching a rolling boil when we got the diagnosis in 2016.

2

Laura decided she needed to tell everyone about the diagnosis at a retreat that the Center for NeuroGenetics had planned for December 2016. The retreat was planned to allow everyone in the Center to get to know each other better and learn about each other's work, and was held at a posh resort hotel on Amelia Island, Florida. A meeting facilitator was present to keep things productive and on track. The group of about 50 or 60 graduate students, post-doctoral students, professors, and clinicians gathered in a hotel conference room to kick off the retreat.

As the Center director, Laura spoke first. She presented PowerPoint slides outlining the goals of the retreat, the Center mission, the expertise of the people in the Center, and the focus of research of the labs. She also described her own background and showed some family photos and then said, "A few of you already know this, but I want you all to know that Bert has been diagnosed with *C9orf72* ALS." She was trying hard to maintain her composure, but tears had already started to fall. "So this work is personal to me." She gave up trying to hold the tears back and continued through sobs. "I plan to do everything within my power to make a difference for Bert and other patients like him."

She was a mess. It had become a catharsis that had been building for years, and through the tears and the struggle to speak in a way that could be understood, she felt some relief from the tightness in her chest. She began to breathe deeply between the sobs, and was able to say, "I hope you will join me in this fight against neurological disease."

Chapter 1:

Florida

The conversation about moving to Florida started when Laura and I were in Costa Rica in January 2009. Laura had organized a scientific conference in San Jose, Costa Rica, and I joined her after the conference for a week of vacation. During a conference dinner one evening before I had arrived, she was chatting with a friend, Dr. Maury Swanson, from the University of Florida. Laura and I grew up, went to college, and had at that time spent our entire lives in Minnesota except for occasional brief vacations or college studies abroad. Like every Minnesotan who escapes in January for a brief but wonderful break from the long, cold, Minnesota winter, she was intoxicated with this new, warm, green paradise and wondering aloud why we lived in a place with such a hostile environment. That prompted Maury to suggest applying for a position at the University of Florida. He said, "We've got a position opening up. Why don't you apply to UF? We'd love to have you."

At the end of the conference, we left San Jose on a de Havilland Otter, a single-engine prop plane seating about 10 people, operated by Sansa Airlines, and flew to Puerto Jimenez, a small town on Costa Rica's Osa Peninsula. The peninsula is famous for the Corcovado National Park, known for having a diversity of biological species not surpassed by a region of similar size anywhere in the world. We landed on a gravel runway just as a goat crossed ahead of the slowing airplane. Our destination was the Bosque del Cabo, an ecotourism resort with luxurious accommodations set amid one of the only old-growth rainforests on the Pacific Coast.

It was a perfect environment for Laura to work on persuading me to consider a move to Florida. The resort is located on a bluff overlooking the Pacific, and the deck of our cabin gave us a wonderful view of blue water far below stretching to the horizon. As we admired the view and watched for Scarlet Macaws and monkeys, the icy bonds that had tied us to that frozen land in the North may have thawed a bit. Still, when Laura mentioned the idea of moving to Florida, I said, "Are you crazy? People go to Florida to die. No one goes there to work."

I added, "And what would I do? You would have to get a big raise to replace my salary if I left Fredrikson."

I had at that time been a lawyer at Fredrikson & Byron, P.A., one of the largest law firms in Minnesota, for twenty-six years. I was a business lawyer with a good practice that had allowed us to send our two kids to private schools and now college. I had some good clients in the Minneapolis area and enjoyed my work. A move would put my practice at risk and, from my point of view, just didn't make sense.

I felt a little guilty about that conclusion, however. I had been a bit of an anchor on Laura's career. Many professors move to advance their careers, but Laura had stayed at the University of Minnesota to allow me to pursue my legal career in Minneapolis. She completed her graduate work there, did her post-doctoral studies there, became an assistant professor there, and then finally became a full-tenured professor at the University of Minnesota. She was successful enough to get occasional questions about whether she would consider a move, but she rejected them all because she knew that moving the family would cause too much disruption.

The possibility of a position at the University of Florida was more interesting to Laura than any of the others for several reasons. She already had productive collaborations with several colleagues there and wanted an opportunity to build a broader research program focused on neurogenetics.

Also, the kids were leaving home. Paul was already away at St. Olaf College in Northfield, Minnesota, and Maddie, two years younger, would graduate from high school in the spring and start at St. Olaf in the fall. A move wouldn't disrupt their lives significantly. We had lived in the same house at 2116 Carter Avenue in St. Paul for their entire lives, so at least in Laura's mind, it was time for a change.

When we returned to frozen Minnesota from that trip to Costa Rica, the seed had been planted. But like most things in the Minnesota winter, it lay dormant for months. When Laura suggested that applying might cause her colleagues at the University of Minnesota to take her less for granted, I thought, *Right, that couldn't hurt.* She worked up an application and sent it in. Maddie didn't like the idea. She said with authority, "Mom, families are like trees. They spread roots. You can't just uproot them and move them to a different state."

I continued to go into the office every day, sit down at my desk, work at my computer, and make and receive telephone calls. But now I observed my day with the possibility of change in my mind. Most of my meetings were internal meetings with other lawyers at Fredrikson about internal firm stuff, not meetings with clients. While they were enjoyable because, after all those years, most of the lawyers at the firm were good friends, it wasn't really important for me to be there. As for client work that actually generated revenue, 90% of that was done by computer and telephone, which I could do, I realized, anywhere.

My most important client, Cardiovascular Systems, Inc. (CSI), held board meetings every quarter that I attended, but the directors flew in for those. Why couldn't I do the same? Even the CEO of CSI lived in California and commuted, as needed, to Minnesota for company business. Although some board members complained about that, they let him get away with it. I began to think it might be possible to stay with Fredrikson & Byron, move to Florida, and travel back to Minnesota as necessary for meetings.

This idea presented a couple of significant challenges. Each state has its own licensing requirements, and a lawyer who practices law in a state in which he is not licensed is subject to enforcement action for the unauthorized practice of law. Some states, like Florida and California, are more aggressive about enforcing their rules on unauthorized practice of law because they see more lawyers from other states moving there and practicing law without a license. I knew I would have to get licensed in Florida in order to practice law there. And I knew to get a license in Florida, I would have to take and pass the Florida Bar Exam.

The other challenge would be generating new clients in Florida. I generally worked with early-stage companies. This was fun because there is an energy in start-up companies that is often missing in large institutional organizations. I worked directly with the founders or management team, and shared the adventure of each successful fundraising or merger or the excitement of an acquisition that provided a liquidating event for investors. I was a counselor to officers and board members, providing business advice as well as legal advice.

But early-stage companies have a life cycle. Some don't make it and often cease business, sometimes leaving Fredrikson with unpaid legal bills. Others, if they are successful, get acquired by a larger company that is represented by other legal counsel, and we lose the client. CSI is the rare exception, a long-term client that I had represented since the founder and CEO came to me in 1994. I had represented CSI through its growth from a start-up company (that almost failed with the market turmoil in 2000) to a public company with a market capitalization in excess of $1 billion. But I knew eventually I would lose CSI as a client also, most likely through an acquisition, I thought then.

Therefore, my continued success as a lawyer required developing new clients. That would be much harder in Florida, where most people would never have heard of Fredrikson & Byron, P.A. In Minnesota, Fredrikson is well known in the business community as one of the top law firms. Clients who hired me knew they were getting a team of top-notch lawyers to work on their matters. I could invite them to visit our impressive offices on the 34th to 40th floors of US Bank Plaza in Minneapolis. In Florida, my business card would have the name of a firm no one had heard of, and I would be working from home, at least at first.

In quiet moments, I thought about what I owed to Laura. Her work, although not as well compensated as mine, was more important. She was breaking new ground in research on neurological diseases, doing work that was one of a kind. I was doing work that could be done by thousands of other lawyers. She deserved more support than the University of Minnesota was giving her. She had shortchanged her career for me and the kids. Maybe I should take a risk with my career for her.

I also considered my age, 51, and the fact that my mother had died from ALS at age 62. We knew from the pattern of inheritance in my mom's family that there appeared to be a 50% chance that I had inherited the gene that caused mom's ALS and also die early from ALS. This made the risk with the career seem insignificant and strengthened my desire to support Laura in whatever she wanted to do. We had been financially conservative and responsible all our lives and now had more resources to take risk with than ever before. And it might be fun, I thought.

By the time that Laura received an invitation from the University of Florida to come down for interviews, I was willing to move ahead with caution.

Chapter 2:

Gainesville

The recruitment process took a year and half. Laura went for the first visit alone in the summer of 2009 to give a seminar and have a round of interviews. That went well, and so I joined her on the second trip in the fall for additional discussions and interviews. Laura presented her list of terms that she would like to see in an offer letter: a start-up package that would allow her to set up her lab, financing for a new Center for NeuroGenetics that Laura would lead, money for additional recruitments to the Center, and a salary with a nice increase from her University of Minnesota salary.

Laura's second visit was my first time in Gainesville. There is no direct flight from Minneapolis to Gainesville. Delta offers several flights each day from Minneapolis to Atlanta with a travel time of about two hours, but then the Gainesville traveler is required to connect to a flight to Gainesville, almost always on the D concourse of the big Atlanta airport, served by one of Delta's partners operating a smaller plane, with a travel time of about an hour. The lack of a direct flight between Minneapolis and Gainesville extended the travel time significantly. With time in the airports prior to flights and connecting in Atlanta, the trip takes about five hours.

Upon landing, we disembarked into the single waiting area of the Gainesville airport, and after a short walk, we were on the street. I was struck by how small the airport was. I told myself, *Small is good. The less time we waste in airports, the better.* But I worried about the size of the market for lawyers like me.

We stayed at the Hilton Hotel on 34th Street, near the UF campus. While Laura was away at meetings, I went for a run. It was hot. I ran through the campus to get a feel for the place. It was beautiful, with red brick buildings along tree-lined streets with a 20-mile-per-hour speed limit. I ran by Lake Alice and smiled at the exotic sight of a beware of alligator sign. I stopped to marvel at the Baughman Center, a curious chapel on the shores of Lake Alice with a gothic-inspired, modern design, with soaring windows. Eventually, I got to Ben Hill Griffin Stadium, aka "The Swamp." As a sports fan, I was aware of the storied history of Florida football and the national championships in 2006 and 2008 and felt as though I was in the presence of greatness. I was surprised to see that the stadium was open, and as I cautiously jogged up the ramp, expecting someone to tell me to get lost, I saw that others were running up the steps inside the stadium. After pausing a moment to admire the field and the empty bowl of the great stadium surrounding it, I too ran some steps in the stadium.

On the way back, drenched in sweat, I thought maybe I could make this work. Gainesville was small, but the University of Florida was an impressive place. I imagined opening a satellite office of Fredrikson & Byron in Gainesville with a few associates. If Laura wanted to come here, I decided, I would work hard to build a practice here.

We came back to Gainesville again in January 2010 to work out the terms of Laura's deal and to look for housing with a real estate agent. At my request, the University set up meetings for me with several leaders in the business community, including David Day, the director of the Office of Technology Licensing, referred to as the OTL. David was 60-something with a full head of well-combed, white hair and an expressive face that he often uses to great comedic effect by matching a serious expression with a hilarious story. I met with him in the OTL office on the UF campus, and he talked about the thriving start-up company environment driven largely by OTL licensing. He described the Innovation Hub, which was planned for construction a few blocks away, and the Sid Martin Biotech Incubator in Alachua, about 10 miles down the road. He said other law firms came in from out of town to work with local companies. He was enthusiastic about the business environment in Gainesville and especially the power of the University OTL to continue to spin off new businesses based on University intellectual property. I walked away from that meeting encouraged about the prospects of getting new clients.

I also met Brian Hutchison at a Starbucks. Brian was the CEO of RTI Biologics, Inc. (now RTI Surgical, Inc.), one such business started with UF technology and now a public company. Brian was probably 50-something and had the friendly, although somewhat reserved, bearing of someone practiced in dealing with people asking for something. We talked about RTI, the business environment in Gainesville, and neighborhoods. I gave him my pitch on Fredrikson & Byron and how I had decades of experience working with companies like his and that I would be right here in Gainesville and happy to help. He explained that they were currently using Fulbright & Jaworski, a well-known New York law firm, but he would keep me in mind. I thought it was a good meeting but knew it would be hard to replace Fulbright.

As the snow melted in Minnesota, Laura continued to communicate via email and phone with the University of Florida and we talked more seriously about moving. I began to look into preparing for and taking the Florida Bar Exam.

The bar exam threatens every lawyer like Scylla and Charybdis threatened Odysseus. To stretch the simile a bit further, let's imagine Scylla, the six-headed monster, is the Multistate Bar Exam, and Charybdis, the whirlpool, is the Florida-specific portion of the exam. Either one can kill the hero and prevent the safe passage of Odysseus back to Ithaca. Anyone who wants to have a prayer of passing has to spend time preparing to navigate both portions of the test.

The exam is two days. Each day is broken into a morning testing block from 9:30 a.m. to 12:30 p.m. and then an afternoon testing block from 2:15 p.m. to 5:15 p.m. The first day is focused on Florida law and consists of essay questions in the morning and multiple-choice questions in the afternoon. The second day is the Multistate Bar Exam with 100 multiple-choice questions in the morning and 100 multiple-choice questions in the afternoon. The Multistate Bar Exam is developed by the National Conference of Bar Examiners and is used by many states to test understanding of legal concepts and legal reasoning.

It is a mistake to assume that an experienced lawyer like me might have some advantage in taking the bar exam. Lawyers typically specialize in particular areas soon after law school. I became a corporate and securities lawyer, and although I was immersed in the minutiae of securities regulation and Minnesota and Delaware corporate law, I had largely forgotten the real estate, constitutional, civil procedure, torts, and Uniform Commercial Code law I had learned in law school 27 years earlier. And I had no knowledge of Florida law. I was at a disadvantage to all those kids who were fresh from law school, especially Florida law schools, and likely to remember more of that information. I also wondered if at 52 I was as able to study and retain large volumes of information the same way I could at 25.

But the truth is, no one relies on what they learned in law school or in practice to take the bar exam. There is an entire industry offering to prepare worried bar candidates for the exam. For a fee, bar review courses plot out the studying and practice exams in a systematic way and provide you with books, video lectures, and mnemonics and other memory tricks, which as a whole give you a good chance of passing the bar. They tout their bar passing rates, which often hover around 90%, but everyone fears being in that bottom 10%.

Intelligence is not enough to pass the bar exam. Among those who have failed various state bar exams are Franklin D. Roosevelt, Hillary Clinton, Michele Obama, and John F. Kennedy, Jr. Although those names gave me the comfort of knowing I would be in good company if I failed, I assumed that failure would diminish the amount of respect I had in the eyes of my law partners, many of whom knew I was taking the Florida Bar. It is a stain on the record of any lawyer, and we all want to avoid it if we can.

When it appeared that Laura was likely to work out an acceptable deal with the University of Florida, I signed up for a Kaplan bar review course for Florida. Kaplan offered online lectures on demand, and so I planned to continue working but try to get home early to do the bar review in the late afternoon and evenings. As recommended by Kaplan, I began studying in May for the July exam.

My plan was to get to the office early, by 7:00 or 7:30 a.m., work until 3:00 or 4:00 p.m., and then go home and study for the bar. I remembered the luxury of studying for the bar after law school when that was all I had to do and it was a full-time job then. I wasn't sure how this was going to work. I soon found that I couldn't get through the full amount of bar review materials designated for each day, even if I stayed up until 11:00 or midnight. I put in extra time on the weekends to catch up and started to go home a little earlier. Soon I was eating breakfast before leaving the house, working at the office until about 1:00, and then taking off for home, where I would grab a bite to eat and start to watch lectures and take notes. I needed about seven or eight hours of sleep, especially with that intense schedule, and so I usually stopped around 11:00 p.m.

Late May through late July was a strange period. I let the compulsive bear in me out of its cage. I stuck to the prescribed schedule. I secretly enjoyed reviewing constitutional law and criminal law again and having all those cases and principles top of mind. I knew I was unlikely ever again to have this breadth of legal knowledge and have memorized so many technical rules. I had trained for the marathon and now it was time to run the race.

We made another trip to Gainesville in June 2010, and Laura accepted UF's offer. It was a good offer including most of what she had asked for. Maddie had concluded that St. Olaf was not for her and was planning to transfer to the University of Pennsylvania in the fall, so she had conceded that sometimes a move is appropriate and necessary.

The bar exam was July 26 and 27 at the Tampa Bay Convention Center. I flew into Tampa from Minneapolis and reviewed my bar review outlines in the airplane. I checked into a hotel near the Convention Center, decided I had done all I could, and had room service bring up something to eat and turned on the TV.

The next morning, I walked to the convention center and joined the crowd of test-takers milling about. Most of them were in their 20s, and I felt conspicuously old. I scanned the crowd and saw a few others who looked a decade or two over 25. We had to go through a metal detector to gain admittance to the testing area. No cell phones were allowed in the testing area. Most people had brought laptops. I had not. I had gone back and forth about whether to bring a laptop to write the essay portion of the exam and ultimately decided to write it by hand. I had done that the first time I took the bar exam, and I wanted to avoid any last-minute technical issues such as a software glitch or a battery failure or some other problem.

There were thousands of us. The testing center was huge, filled with tables and chairs. Two test takers sat on the same side of each table facing the front of the room. Each of us had been assigned a number, and we found the table corresponding to our number. I was nervous. A proctor at the front of the room gave us instructions. We were not to open the folder with the questions that were being distributed to each table until he told us to begin. When the time was up, we were to immediately stop typing or writing. A digital clock was visible in the front of the room showing how much time remained.

He announced we could begin. I read the questions briefly, made a mental note of how to allocate the three hours, then read the first question carefully. Recognizing several issues to discuss, I began writing. By the time I started the second question, I wondered whether I had made a mistake in deciding to write by hand instead of using a laptop. I wasn't able to write as fluidly and quickly as I remembered. By the end of the first hour, my hand was getting tired and my handwriting began to deteriorate. I worried about whether the graders would be able to read it. I pressed on.

During the third hour, I started to feel an urge to pee. With my slow handwriting, every minute was important, but so was the need to pee. I fought to maintain focus on answering the question, then looked at the clock and recognized I was going to have to pee before the session ended. So I had no choice. I got up, and though I wanted to run, I forced myself to walk the long walk to the restroom in the corner of the cavernous room. When I returned, I redoubled my efforts at answering the questions and making my hand move as quickly as possible to produce legible handwriting. The time mercifully expired on the essay portion shortly thereafter.

During the exodus for lunch, I tried not to listen to the kids talking about the hidden issue in the third question, or their answer, different than mine, to the second question. I didn't feel bad about my answers, but I didn't feel good either. With tests like these, I knew that my mind tended to gravitate to those areas about which I was uncertain and discounted the stuff I knew cold and could easily discuss. I ate my lunch and steeled myself for the afternoon session. I limited the amount of liquid I drank at lunch and made sure to empty my bladder before the afternoon session.

The afternoon session covered Florida law in a multiple-choice format, so there were no handwriting issues and my precautions to avoid a restroom break were successful. I used all of the three hours and finished most of the exam. In the final minutes, I may have filled in several circles on the answer sheet without reading the questions in the hope of getting one or two correct answers by luck.

The second day was less stressful. I knew the drill, the layout of the place, where the restrooms were, and that I could easily blacken the circles of the answer sheet with my No. 2 pencil. I knew also that the enemy on the multistate bar exam is time and that I would have to move quickly. The three-hour morning and afternoon sessions each contained 100 questions, and we knew that we had to do about 33 questions per hour and about 17 each half hour. That means that if you spent two minutes on a question, you were going too slow. All that I remember about the multistate exam is trying to maintain focus and read quickly. It is a test of how well you can maintain focus over a three-hour period as well as your knowledge of the law.

I felt hugely relieved when it was over. I called Laura from the hotel room after that second day and said, "I don't know if I passed, but I gave it my best shot." The next day I flew to Dallas, where I met Laura and Maddie at the airport. We flew to Costa Rica for a 10-day vacation as a reward for having survived this stage of the odyssey of our move to Florida.

Chapter 3:

My Career as a Salesman

Most law students don't appreciate the business of law when they go to law school. I certainly didn't. I assumed clients just appeared at the law office and the law student's job was to learn the law and how to represent those clients professionally and ethically. As a young lawyer working at a law firm, I began to realize that senior lawyers had some kind of relationship with clients, but I was so busy doing the work that the senior lawyers assigned to me that it seemed to me that not being busy was highly unlikely and, if it did occur, would be welcome. The firm management talked to us about marketing, community involvement, and client development, but I really learned the lesson from watching the dynamics among lawyers in the firm and looking at the numbers that our firm distributed every month.

Fredrikson & Byron, P.A. has a very transparent management culture. Every shareholder receives every month a report that lists the hours that every lawyer has worked on billable client matters (which often are significantly less than the total number of hours that lawyer has worked) and at the end of the year a spreadsheet that shows the amount of fees charged to that lawyer's clients and the compensation of each lawyer. Everyone knew the score, including who was busy, who had a big book of business, and who did not. Success was equated with being busy, having a big book of business, and making a lot of money. The key metrics determining how much money you made were the number of hours you billed overall, whether to your clients or other lawyers' clients, and the number of hours billed by any lawyer to your clients. It is a very quantitative system.

Young lawyers are busy because senior lawyers like to provide work to young lawyers to train them and because their billing rates are lower. Senior lawyers also like to give young lawyers mundane work that few enjoy, like drafting routine documents or reading voluminous material for relevant issues, that does not demand the special expertise of a senior lawyer. Senior lawyers appreciate the clear line of authority in which the junior lawyer is subordinate and subject to the sometimes arbitrary demands of the senior lawyer. As a lawyer advances in experience and his or her hourly rate goes up, however, senior lawyers are more cautious about what work they assign to that lawyer, and the advancing lawyer is expected to generate clients of his or her own. Senior lawyers with their own clients are independent and respected by their peers. Senior lawyers without clients of their own are dependent on other lawyers providing work to them and are lower on the totem pole.

Therefore, if you want to be successful in law, you must be effective at getting new clients or lucky enough to inherit them through the firm. Since I would not be in the office day to day, I would be unlikely to be pulled into teams on which I could bill time to other lawyers' clients or inherit clients. Alone in Florida, I would be dependent on maintaining relationships with my existing Minnesota clients, and I would have to develop new clients. I would have to be a salesman.

I don't mean to suggest that being a salesman is negative in any way. I admire those who are skilled at sales, just as I admire talented musicians or athletes. Good salespeople have a talent for interpersonal relationships. They have a sensitivity and an empathy for other people. They are good communicators. They are often good-looking, and they understand the theatrical aspect of presentations. Most importantly, they are able to inspire trust and win the confidence of potential clients.

I was fortunate as a young lawyer to work with a wonderful mentor, Tom King, who was a talented lawyer good at client development. In college, Tom was a running back for the Minnesota Gophers football team when the team went the Rose Bowl in the early '60s, then went on to law school at the University of Minnesota. He had the attitude of an athlete, ready to take on a challenge, find the fun in working hard, and he had a sportsman's sense of fair play. He was over six feet with a square jaw and gray hair even in his forties when I first met him. He looked like a lawyer out of a movie set and was the kind of guy people liked to be with.

I worked with Tom for a couple of decades and learned that Tom's skill at client development went deeper. He had a talent for rapidly distilling complex problems to their key points and presenting those points to the relevant decision-makers. By contrast, many lawyers bury clients with so much detail that the key points are hard for the client to identify. Tom also never lost sight of how decisions would affect people and the fairness of those results. Therefore, he was widely trusted for his judgment on sensitive, difficult issues, the kind of decisions where everyone is unhappy no matter what the result. He served on Fredrikson's board of directors and was the chairman of the board until he resigned when he concluded it was time to make room for younger lawyers.

When I told Tom the whole story about Laura's interest in going to Florida and reviewed the pros and cons of what it would do to my career, he said simply, "Well, you really have to go, don't you?" That's what I mean when I say he was good at distilling complex problems.

While it was my great good fortune to work with Tom, I also realized that I was no Tom King. I knew that it would be hard for me to develop clients in Florida despite all my advantages as a white male. I rationalized that client development is a skill that can be learned and that depends more upon proper execution of strategies than personal charisma. I think I've heard that in one of the client development seminars I've seen.

I was starting from a good place. My client CSI produced good billings every year, and we had a team of lawyers serving them. In addition, I had a handful of other clients that provided sporadic work. I thought I could maintain my Minnesota practice by traveling from Gainesville to Minneapolis for a week every month. The other three weeks I would work remotely from Florida and work on marketing to potential Florida clients.

Another piece of the puzzle fell into place when I received the news that I had passed the bar exam. The news came by email a couple of months after the exam. It was anticlimactic. I was mostly relieved that I hadn't failed rather than elated that I had passed. My emotional barometer was tied to my expectation of passing, and so although failing would have been a significant blow, passing didn't give me the lift I had expected. Still, I was satisfied that all that work had paid off.

We decided to keep the house in St. Paul, at least for the time being, for my monthly return visits. I also felt more comfortable keeping the house until we were sure Laura had settled in and was happy to stay. It was an escape valve in case Florida didn't work out.

Finding a house in Gainesville was challenging. During every visit to Gainesville, we had driven to various neighborhoods with our realtor, an engaging and friendly fellow named Henry Rabell. After we had looked at what I assume was every listing in our price range in Gainesville, Henry had told us, "Maybe you should think about building a house." We hadn't found a place when Laura accepted the offer in June 2010, and we began to think about renting something until we knew the area better.

We flew back to Gainesville in September to continue our search for housing. Laura found a house online on Zillow whose listing had expired with big porches on the front and back that attracted her attention. We had Henry call and ask to see it, and although the owners had taken it off the market and initially said it was not ready to show, Henry explained we were leaving town the next day, and they relented. In addition to the lovely porches, the house was located on a lot of almost two acres with many trees, a contrast to our small urban lot in St. Paul with houses on each side within 10 or 20 feet. Built in a gated neighborhood in 1994, it also had a lot of windows and a big Florida great room. There were some problems, though. The wood floor had some cupping of the floorboards, indicating too much moisture in the crawlspace.

We liked it enough to make an offer with a significant deduction for the potential floor problem and subject to inspection. We reduced the price even more when the inspection came back indicating we would have to remediate the crawl space, install a dehumidifier, and replace the wood floor. We were told the sellers were unhappy, but they agreed to move forward with the sale. We made one more trip to Gainesville for closing on October 22, 2010.

We planned to leave Minnesota in the morning on Saturday, November 8, 2010, but instead left early on Friday afternoon when the forecast showed a blizzard arriving Friday evening. We packed up our SUV with clothing, personal items, and everything we thought we needed to establish a household. We left St. Paul after it had started snowing and drove well into Iowa, south of the bad weather, where we stopped for the night. When daylight arrived, the leafless oaks stood like sentinels against an overcast sky. The black pavement and highway, the snow, the trees, and the gray sky created a cold, black and white, winter tableau. We drove through the seasons in reverse as we went south. By the time we reached the hills of Tennessee, we had returned to autumn with sunlit hues of red and gold spread across the hills. By the Georgia/Florida line, we were back to summer. It was warm and green, and we needed our sunglasses as we drove. Our spirits brightened along with the weather.

At the house in Gainesville, we camped in the bedrooms above the garage while various contractors worked on the crawlspace and the main floor. The project grew as we evaluated the kitchen, decided on converting a bedroom to my office, and spoke to a designer, Elyse Ostlund, whose advice we came to trust and respect. We caught the "as long as you're at it" syndrome that affects so many remodelers. In addition to encapsulating and drying out the crawlspace and replacing the wood floor on the entire first level, and then finishing and sanding the new hickory floor, we put in a new kitchen, built a desk and bookshelves in the office, and added a new entertainment center in the great room. Then we had the entire interior repainted. Some of that may have been desperation to make this a comfortable home in a new place that didn't feel like home.

I started to work remotely from my new home office, energized with the challenge of building a practice in Gainesville. Laura was also working hard to establish herself in her new position, and we both rose early and worked late. Every month I flew back to Minnesota for a week, initially arriving on Sunday and departing on Saturday so that I could spend a week with full days on Monday and Friday in the office. I stayed alone in the house at 2116 Carter Avenue in St. Paul.

Laura and I say now that time slowed down during those early years after the move to Gainesville. In our recollection, the first two years in Gainesville were about equal to the previous 10 in St. Paul. Everything was new and different, and our experience of life had less of the mind-numbing routine that develops when you've lived in a place for 30 years. We also say that it took about two years to really feel at home in Gainesville, so the price of that stimulating experience of a new place is a little discomfort and the stress of a new job challenge.

I discovered that Gainesville really is a small town, and the business community and the University of Florida are key players in an ecosystem that is pretty good at growing businesses. I mentioned earlier that I met David Day, director of the UF OTL, and Brian Hutchison, CEO of RTI Surgical, Inc. (RTI), on one of our visits to Gainesville. David introduced me to Jamie Grooms, one of the co-founders of RTI and later a serial entrepreneur and overall mover and shaker in the Gainesville business community. Jamie introduced me to Richard Allen, another co-founder of RTI and serial entrepreneur and shaker and mover. Their experience with RTI evidently gave Jamie and Richard the appetite, and probably the capital as well, to start more companies. These second-generation companies include AxoGen, Inc., co-founded by Jamie and Richard, and Xhale, Inc., founded by Richard. I met everyone and let them know that I was interested in taking on their legal work.

Another example of the surprising connections in a small town relates to Applied Genetics Technology Corporation (AGTC). The UF professor responsible for sparking Laura's initial conversations with UF, Maury Swanson, is married to Sue Washer, the CEO of AGTC. Since Maury and Sue had dined with us at various points in the recruiting process, Laura and I considered them friends, and shortly after we arrived in Gainesville, I scheduled lunch with Sue to talk about Fredrikson and AGTC.

Of course, these companies already had law firms working for them, and it is difficult to replace an existing relationship. So, although I was able to generate a modest amount of legal work from the established Gainesville companies, I was largely unsuccessful at getting any significant projects. I've tried to parse out the reasons my high hopes were left unrealized, and I end up grasping at metaphors, like baseball. I'm not a slugger, but I've thought of myself as a veteran hitter who would be able to hit some singles and doubles. The trouble with Gainesville is that the number of times you get up to bat is limited. There just aren't that many potential clients. Or, to use the wolf pack metaphor, I had left the pack, and because I wasn't sharing in the big game that the pack brought down anymore, I was starving.

When it appeared that I had struck out with the obvious potential clients in Gainesville, I resolved to settle in for the long term. I joined the local chapter of BioFlorida, which bills itself as the "voice of Florida's life science industry." BioFlorida puts on a big annual convention that attracts companies like RTI, Xhale, AxoGen, and AGTC from all over the state. Between annual conventions, the local chapters put on smaller events for their regions.

Our job as a local chapter was to organize about three events per year to allow people in the life science industry to come together, network, and hopefully learn something. I volunteered to help and found that finding and scheduling speakers, a venue, and caterers and following up on the innumerable details, including projectors, mics, tables, and chairs, was a lot of work. But the advantage of being an organizer is that I could cast myself in a prominent role such as moderator and surround myself with important people before an audience of people in the life science industry.

I volunteered with BioFlorida for several years, eventually becoming chairman of the local chapter. I was able to get Fredrikson to sponsor several of these meetings, in addition to paying the annual BioFlorida membership fee.

I remember one meeting in particular on February 14, 2013. The meeting was about "Equity Financing for Life Science Companies." I invited Richard Allen, Sue Washer, and an officer of AxoGen, Greg Freitag, to join me as panelists. I moderated. We held it at the same Hilton Hotel on 34th Street where we had stayed on my first visit to Gainesville. About 40 people attended. Richard, Sue, and Greg each talked about their experiences in getting financing for their companies, and I talked about the legal issues involved in selling securities.

I had spoken enough in the past to know when a talk is going well, when I felt like I was thinking aloud and words flowed directly from my brain to the air, articulate, clear, and intelligent. In these situations, the mouth and throat are so subordinated to the overwhelming force of the idea, and their operation so quick and flawless at communicating the idea, that they don't seem to be involved. I did not have that experience this evening. My talk seemed laborious and wooden. The topic, securities law exemptions, didn't help. I was glad when it was over and we turned to the refreshments and socializing.

As I walked to my car in the dark parking lot after the event, the energy of being in front of a group drained away. My part in the presentation had been terrible, I thought. As I drove toward home, a tightness rose in my chest. I was failing at this effort to build a practice in Gainesville. These BioFlorida meetings were useless. No one in that room that night would hire me as a lawyer. Tears came and I swallowed hard. There would be no Fredrikson office in Gainesville.

My mind was a descending spiral of negative thoughts. Not getting clients would lead to lower billable hours and lower income. The firm might conclude that working remotely was unsustainable if I didn't produce revenue. Despite all my high hopes and hard work, I was failing. Then I turned to my increasing worry that the family history of ALS was catching up with me. *You're worried about your career?* I asked myself. *What about ALS?* I suddenly realized that the difficulty with the talk I had given might be connected to the increasing number of incidents in which I had become aware of diminished speaking abilities, which I knew was an early symptom of ALS. I told myself then, even though I would not be diagnosed for several years, *The laborious and wooden talk. You have ALS, and it is progressing.*

When I got home, and Laura asked how the event went, I said, "Fine."

Chapter 4:

Confronting the Monster

I first noticed changes in my voice in 2012. I was sitting in my office at the law firm in Minneapolis talking on the phone, and a tiny glitch in pronouncing a word (I don't remember the word) caught my attention. I'm sure the others on the call did not notice it, but after hanging up I said the word over and over, trying to decide if this was just a one-time thing or the beginning of something more serious.

Later, probably the same year, we were at Lake Itasca in Minnesota for a family reunion. I was standing by the campfire talking about something, when Sandy, a nurse practitioner, said to me with a smile, "Bert, there's something about the cadence of your speech that's changed. You've developed a little southern drawl since you've moved to Florida!"

I thought, *This is the first time anyone has noticed a change in my voice!* But I said in an exaggerated southern drawl, "Why, Shugah, how kind of you to notice mah efforts to speak like the locals." The conversation moved on from there, but the memory stayed with me.

I kept it to myself for over a year to savor the days of being "normal" and also because I knew Laura would take it hard. Even after the meeting in 2013, I didn't want to say anything for fear of upsetting Laura. Eventually I told her that I had something going on with my voice and was worried about ALS. Laura is a sensitive and empathetic soul, and we hugged and cried and then talked. I said something like, "We have to avoid fear and negativity. They do nothing to address the problem. Let's focus on the present. We are both healthy otherwise and able to enjoy life, and we should take full advantage of our many blessings. If we get mired in self-pity and fear of what may happen down the road, we'll ruin the present. We will all die, but we can't live with joy if we are constantly afraid of that death." I believed intellectually that I could banish fear, but my gut was not so sure.

The fear that dominated all others for both Laura and me was this thought: *If I have the ALS gene, there is a 50% chance that Paul and Maddie have the gene.*

We decided to see some neurologists at the University of Florida. They examined me, gave me an MRI and an EMG, and said they saw nothing wrong. I took that as good news. Still, my voice was changing, and I knew from the internet that about 25% of ALS cases begin with the voice and swallowing. I thought back to my handwriting issues on the bar exam and suspected that my handwriting had started to change even back then in 2010. We told the kids, who said they had been already wondering about my voice. Soon other people were asking about my voice too.

In October 2015, Laura, Paul, Maddie, and I all flew to San Francisco for the wedding of Nelson Whitmore, our nephew and Paul and Maddie's cousin, to Camilla Danpour. We drove to Camilla's hometown of Healdsburg, where the wedding took place on Saturday. On Sunday morning we joined the wedding party for a brunch and spent the afternoon touring the wine country. We stopped at the Korbel Winery, established 1882, and had a tour, then went to Armstrong Woods state park, where we walked in awe among the redwoods. The sign by the Armstrong tree said that the tree was about 1,400 years old.

In the evening we had a lovely dinner at Baci, which had been recommended by someone we met in the hot tub at the hotel. On the way back to the hotel, as we approached an intersection, Laura, with GPS on her iPhone, suddenly said, "Turn left here!" As I was nearly in the intersection, but not in the left turn lane, I stopped and waited for the cars in the left turn lane to make their turns, thinking I could follow behind them. But by the time the last one went by, the light had turned, so I edged the car over, about halfway in the left turn lane, turned my blinker on, and waited for the light.

A policeman drove up behind me. He got out, walked up to my window, and asked what I was trying to do, then asked me to go through the intersection and stop ahead. When we were safely at the curb on the other side of the intersection, he walked up and said, "Could I see your driver's license, sir? Where are you coming from tonight, sir?"

"We were at the Baci restaurant for dinner—just going back to the hotel," I said.

"Did you have something to drink at the restaurant?"

"I had a glass of wine with dinner."

"Just one?"

"Yes, just one."

"Do you have a speech impediment, sir?"

Of course, I knew he was thinking my speech was affected by alcohol, and I immediately concluded that I should say "yes" to disabuse him of that notion. But, I asked myself, do I really have a speech impediment? My identity was tied up so tightly with speaking well and being articulate that it was hard to admit that to myself, especially there in front of my family, all listening intently to the conversation. However, I concluded that I needed to accept reality and said "yes."

Then he went back to his squad car to run my license. When he came back, he asked me to step out of the car and join him on the sidewalk. I got out of the car and onto the sidewalk, feeling like my family in the car and even the people in the cars passing by were staring at me.

"I am going to hold up my index finger like this and move it back and forth, like this. I'd like you to follow it with your eyes, while holding your head steady, straight ahead. Can you do that?"

"Yes," I said, confident that I could perform the required task. He moved his finger, and I tracked it with my eyes without moving my head.

He said I could go. I returned to the car a little embarrassed to have disrupted what was otherwise a lovely day.

In February 2016, I got a call from Kathy Petrucci, the CEO of a start-up company in the San Diego area developing stem cell treatments for the veterinary market, Animal Cell Therapies (ACT). Kathy is the daughter of Gary Petrucci, a savvy financial player in Minneapolis who saved CSI several times and gave it a chance to succeed. Gary got me involved as counsel for ACT in 2008, and I've known both Gary and Kathy for years and consider them both friends as well as clients.

ACT's stem cell treatment was used with a small dog, Buffy, who stars in a remarkable before and after video. In the "before" video, Buffy is shown dragging her hindquarters across the floor, one of her rear legs completely useless. In the "after" video, Buffy is shown walking slowly outside, although with a limp, and climbing three steps to the door of a house. I understand that Gary showed the video to a contact that he has at the Mayo Clinic, and subsequently Mayo asked for a conversation with ACT. Kathy, Gary, and I, along with others involved with ACT, had a call to prepare for the conversation with Mayo.

After the conference call, Kathy called me and said, "Bert, are you okay?" Kathy is probably about five foot three, and when I saw her last years ago, she had shoulder-length brown hair and a serious business-like expression that would give way to a energetic smile when called for.

I like Kathy's directness. I said, "Well, I've got this voice thing going on. I'm sure that's what you're talking about."

"Have you had a stroke?"

"No, no," I paused, wondering how far to go. "I've been worked up by the neurologists, and they don't really know what's going on." Which was true. They could point to nothing wrong.

"Is it degenerative? Do you think it's getting worse?"

"It's been going on for a long time. I don't know if it's getting worse, maybe, but I'm trying to do everything I can to stay healthy, getting enough sleep, eating right. And I feel good, except for my voice. Getting old, I guess."

"But you're not that old."

"Right, I agree! Not old enough to sound like this." A pause as I thought, why not just put it on the table? "We're worried about ALS, which often begins with speaking problems."

"Do you have any muscle problems?"

"No, I feel great; I have no muscle issues."

"Is it genetic? Is there some genetic connection?"

"Well, that's why we're concerned. My mom died of ALS."

"We should get you into a clinical trial for stem cells. I'm going to look around and find out what clinical trials are taking place."

"That's very kind of you, Kathy," I said. "We're pretty up to date on the latest stuff. My wife is a researcher on neurodegenerative diseases, and so she knows the field; she's all over this. But it's thoughtful of you to offer."

"I'm going to look around, do some research. I like you, Bert. We've known you for a long time. I'm sure there are clinical trials using stem cells to treat ALS. You won't mind if I send you stuff, will you?"

"You're very sweet, Kathy, and I like you too. Sure, send me whatever you find—I am always willing to look at anything that might be helpful. But don't worry about me. Really, I feel good. I'm optimistic and hopeful about things. The medical progress made every day is amazing. Who knows, there may be a treatment tomorrow that can fix everything. But thanks again."

About 10 minutes after the call, Kathy sent me an email with a link to a YouTube video of two Mayo docs talking about their clinical trial using stem cells to treat ALS. In the text she said, "I'll ask them about this on Friday. Maybe this is meant to be and the real purpose of the call on Friday."

Chapter 5:

Diagnosis

As my voice continued to deteriorate in 2016, Laura made an appointment with Dr. Jeffrey D. Rothstein, MD, PhD, Director of the Robert Packard Center for ALS Research and Professor of Neurology at Johns Hopkins University. I was reluctant to travel all the way to Baltimore to see a doctor who couldn't really do anything for me, as I thought then, but I ultimately agreed.

Laura has known Jeff for years as a result of a collaboration they had on another disease called spinocerebellar ataxia type 5. Jeff has a ready smile and looks too young to have accomplished so much. He talks in a calm but rapid manner that suggests he is a smart man in a hurry. I first met Jeff and his wife Lynn at a scientific meeting in Costa Rica that Laura had organized (yes, another one) and to which I tagged along just for fun. Lynn and I arrived at the Costa Rica airport near the meeting at about the same time and shared a car up the mountain to the rustic resort, where the meeting had begun a few days earlier. Lynn is also a lawyer and worked in the U.S. Office of Patents and Trademarks, and so we chatted about patent law and start-ups.

We arrived in Baltimore on April 24, 2016, and after checking into our hotel, Jeff picked us up and brought us to his home in suburban Baltimore for dinner. Laura and I were surprised and pleased by the unexpected and gracious invitation. Jeff and Lynn had built a lovely, traditional, brick house a few years before, and we enjoyed a casual dinner in their beautiful kitchen. After dinner, Laura and I walked through the neighborhood with Jeff while Lynn attended to some pressing, work-related matter.

I asked Jeff how he had seen ALS progress in patients with initial symptoms involving the voice. I had read that ALS often progresses more rapidly when symptoms begin with the voice or swallowing, referred to as "bulbar onset" cases. In most such cases, death follows three to five years following diagnosis. My mother's ALS had begun with her arms, and the time from onset to death was less than a year. But Jeff mentioned that some cases of ALS do not progress so rapidly. He said that he had one patient whom he had been seeing for about 10 years. I was encouraged by his comments.

The next day, at the outpatient clinic, Laura and I were introduced to the electronic registration system at Johns Hopkins. As we arrived at the fifth floor where the Neurology group was located, signs directed us to a row of kiosks, each of which had a screen and inputs for identification or credit cards. I felt like I was at the airport checking in at the self-check-in kiosks. A helpful attendant was standing by to offer assistance if a new patient got lost in navigating through the system. We carefully read the instructions and tried without success to get the machine to scan our insurance card. The attendant saw us repeating the process and stepped up to help. We ended up having to key everything in and finally completed the registration process. Then we were directed to a device that prints out a ticket with a number and took seats while waiting for our number to be called.

Somewhere in the process I saw the name "Epic," which I recognized as one of the big medical records software companies. I recognized it because our very smart nephew, Charlie, got a job with Epic in Madison, Wisconsin, upon graduating from St. Olaf. I wondered if Charlie had designed some of the stuff that was now managing my information. Only later did I realize that the system gave me access to my information via my iPhone and computer via an app called "My Chart."

Soon our number was called, and we walked up to the counter station over which our number was displayed. The attendant there gave us a blue sheet and told us to give it to the nurse. Then we sat down to wait some more. Then a nurse came out and called my name and brought me to a small room, told me to step on a scale, noted my weight (155), and took my blood pressure (110/70), heart rate (61), and temperature (98.7), all referred to as my "vitals," as in "Did you get his vitals?" We returned to the waiting room to wait some more. Eventually another nurse came out, called my name, and brought us back to an exam room and told us to wait. After a few minutes, Jeff appeared and greeted us warmly.

We talked for what I'm sure was longer than the insurance-approved time, and then Jeff examined me. The examination included use of a little hammer to test my reflexes and a request to follow his finger with my eyes, without moving my head, as he moved it back and forth and up and down in front of me. I recalled the police officer doing the same thing on the side of a street in California. He also performed various strength tests that involved either pushing against Jeff's hands with my arms, legs, shoulders, or feet or trying to resist Jeff's push against various body parts. I recalled all my years of exercise and the quantification involved in assessing fitness from every pound lifted to every tenth of a second in a race. The neuro exam seemed very subjective by comparison.

Jeff said that although bulbar onset ALS begins with upper motor neurons that enervate the muscles related to speech and swallowing, it also often increases the spasticity of movements controlled by the lower motor neuron. Jeff said that he could detect a "spastic catch" in my arm. My heart sank a little on hearing this because I had been telling myself, "At least so far my arms and legs aren't affected." So, I thought, the slowness of my handwriting that had been developing over the past few years probably wasn't just due to age or lack of practice.

During the exam, Jeff mentioned that there was a clinical trial in the pipeline for a new ALS drug from Biogen that might offer some promise. Jeff and Laura talked about the science behind the drug, calling it an ASO for the *C9orf72* mutation, the most common genetic cause of familial ALS. We knew from Laura's research that my mother had the C9 mutation and had assumed once my symptoms began that I had inherited the gene.

Jeff sent me down the hall to another doctor who performed an electromyogram (EMG). The doctor attached electrodes at various spots on my arms and legs and then attached wires to the electrodes that led back to a machine with mysterious gauges and switches.

He explained, "We're going to send a mild electrical impulse through key nerves in your arms and legs and record the speed at which the impulse travels through the nerves. We can then evaluate the results to determine the health of the nerves." I recalled the occasional encounters with electric fencing in my youth, and the jolt that resulted from contact with the wire. I braced myself for something similar. Fortunately, the impulse was indeed mild and quite tolerable.

A second part of the test involved a more unpleasant insertion of needles in various muscles, after which the doctor told me to move an arm or a leg to cause a contraction of the muscle. The EMG records the electrical activity of the muscle when contracted and at rest, and the data is then evaluated to determine whether there is abnormal activity. The doctor didn't give us any results immediately upon completing the test, and we assumed the data had to be interpreted.

At some point during the day, I urinated in a cup and gave it to a nurse and sat while a nurse drew several vials of blood from my arm for various tests. The most important test would determine whether I carried a mutation in the *C9orf72* gene. My blood was sent to a diagnostic testing laboratory, Athena Diagnostics, to perform this specialized genetic testing.

At the end of the day, as we started on our trip home, I realized that no one had told me that I had ALS. We were all assuming it, but no one had expressly stated a diagnosis. Maybe it was the lawyer in me, but it seemed a conspicuous absence. There was no dramatic moment of the type that we see often on TV where the doctor tells the unsuspecting patient and spouse, "You have ALS." After which the patient and spouse look shocked and tears are shed. Maybe the jury was still out. Maybe the C9 test would come back negative for the mutation, and we would conclude my problem was some as-yet unidentified voice problem. Hope springs eternal.

A few days later, I was exploring the "My Chart" app on my phone, and my pulse quickened when I saw an icon for "test results" and then an item for "C9ORF72 DNA TEST." I clicked on that and read "POSITIVE— this test identified a pathogenic mutation in the *C9orf72* gene." I read it a couple of times and then dismissed it with the thought, *Well, we already knew that.* Then I clicked on the icon for "Heath Summary" and "Health Issues" and read "Dysarthria—noted on 4/25/16" and "ALS (amyotrophic lateral sclerosis)—noted on 4/25/16." I took a deep breath and exhaled. There it was, in writing. And I recalled all the survival estimates expressed in years from diagnosis. Now I had a diagnosis, and I imagined a clock starting. How many years from diagnosis did I have? Three, five, 10?

I made a lot of changes in 2015 and 2016 after confronting and accepting my ALS. I let go of my career goals. I stopped traveling back to Minnesota every month. My clients slowly disappeared. CSI hired the excellent young lawyer that I had assigned to work on CSI matters years before. He became general counsel, and my work rapidly decreased. I assigned the billing responsibility for CSI to an excellent young female lawyer at Fredrikson who had done a lot of work for CSI.

I lost another client, Galil Medical, Ltd., when it was purchased in 2016 for $84,500,000 (with an additional $25,500,000 payable if certain milestones were met). Clients like that have the right to lawyers with effective communication skills, and in a moment of frustration, I apologized to the CEO of Galil, Marty Emerson, for my slow speech. Marty graciously said that although he had noticed the change in my voice, he didn't think it appropriate to ask about it and didn't think that it mattered, as he didn't see any adverse change in my work. As Marty's comments indicate, he is a thoughtful and generous soul, as well as a great CEO.

The changes improved my life. I felt less stress. I got more sleep and exercise. I began to read and write more. I have tried to embrace gratitude and think more about how to live the remaining life that I have. I was surprised by some unexpected events that softened the blow of my declining career. I made a claim under my disability insurance, which was accepted, and I started to receive disability payments that helped to offset my dwindling income. And my confidence was given a boost when Brian Hutchison, CEO of RTI Surgical, Inc., offered me a job as general counsel of RTI in the summer of 2016. I declined the job because of my health issues but felt good about having been asked.

We decided to sell the house in St. Paul. Then Paul and Susan announced their engagement. We coordinated the sale of the house to make sure it occurred after any wedding-related events. Before either the sale or the wedding, however, we heard exciting news about the clinical trial for the new ALS drug.

Chapter 6:

Screening

Over Labor Day weekend in 2018, we went back to Minnesota to begin the process of removing the 30 years of stuff that we had accumulated in our house on Carter Avenue in St. Paul. We were on our way back to Gainesville and waiting in the Atlanta airport for our connecting flight to Gainesville when Laura got an email from Jeff. It arrived at 12:09 p.m.:

"We should talk. Biogen C9 ASO now ready to enroll."

Jeff's email forwarded an email to him and the clinical study team at Johns Hopkins from a person at IQVIA, a contract research organization, or CRO, that was working with Biogen on the study. That email displayed a time of 12:01 p.m.:

"Dear Dr. Rothstein and Kristen: I am pleased to announce that all site activation requirements on the Biogen 245AS101/C9orf72 have been met and you may now begin screening and enrolling in the study…"

Laura's response by email was at 12:33:

> "Oh this is GREAT NEWS!!!! We are about to board a plane to Gainesville but would be available to talk now from airport or after about 2:30."

Jeff at 1:01:

> "Let's talk at 2:30 or so—I leave for Montana around then (so maybe a call from the airport)"

Laura responded at 1:06 with her phone number and then again at 1:40:

> "Our flight is late taking off and now predicted to land about 2:40 pm. If I do not pick up we are still in the air and I will call you back when we land. If that doesn't work, any time after we land today or tonight works for us."

Jeff at 1:46:

> "Ok
> I board at 3:30 and don't get to Bozeman until 8:30 tonight—their time
> I need to review the trial details—but if Bert wants to be considered for the first dosing (5mg) than we/you need to let Lora and Kristen know (I have ccd them). There will eventually be higher dose cohort—but Biogen has to wait to get FDA approval for those higher doses (the USA sites are not

currently allowed to go to the higher doses, Europe can—but Europe may not start until late this year or early 2019—and we expect FDA to remove the higher dosing hold in the next month or so."

Laura at 2:33:

"We just landed will be off plane and can talk in 10 min. Will that work for you?"

No response from Jeff. Laura at 2:52:

"Lora and Kirsten,
Bert is interested in participating please let us know what he needs to do to be considered.

Jeff,
Please give us a call when u get a chance. Cell is good number xxx xxx-xxxx.

Thanks so much, Laura"

Kristen Riley, a member of Jeff's team, responded at 3:22:

"Hi Laura,
Are you available next week on Tuesday or Wednesday to come to Johns Hopkins to screen?

Also, can you send me Bert's full name and date of birth so I can mark him as a candidate for the study? Thanks very much,
Kristen"

Laura responded at 3:26:

"Kristen,
Yes, we will make that work—please let us know what time of day the appointment will be so we can book flights.
Best, Laura"

Later that evening we had reservations with Delta to fly to Baltimore and with a hotel for the overnight stay. Flight schedules make it impossible to get to Baltimore, have the appointment, and get back the same day.

We were able to act quickly when the opportunity arose because Laura and I had discussed it thoroughly and given it a lot of thought earlier. Signing up to have an experimental drug injected into your spine is not something one does casually. But Laura lives in this world. She knows how the drug works. She knows Jeff.

She even knows the company that developed the drug, Ionis Pharmaceuticals, because she collaborated with them to test a similar drug for the disease. A few years ago, I accompanied Laura on a trip to Ionis in Carlsbad, California, and we had dinner with Frank Bennet, one of the founders of Ionis. At that time the company was called Isis Pharmaceuticals, but they wisely changed the name after the rise of the militant group ISIS in the Middle East. Frank impressed me as a smart, thoughtful man, and so the company already had credibility with me before I knew that they had a drug for C9 ALS. Ionis is collaborating with Biogen on the study.

Ionis's platform technology is called "antisense" therapy. This technology is designed to destroy specific RNAs and thereby prevent problems that might be caused by those RNAs. The ALS in my family is caused by a mutation in the *chromosome 9 open reading frame 72* gene (often referred to as *C9orf72*). This mutation consists of six DNA nucleotides that are repeated over and over again, like some printer malfunction printing GGGGCC hundreds to thousands of times. The G stands for the nucleotide Guanine and the C stands for the nucleotide Cytosine. Unaffected individuals might have 10 or 15 repeats, but individuals at risk for the disease often have 400 to more than 1,000 repeats. Scientists call this type of repeat a "hexanucleotide repeat," and other diseases are caused by repeat mutations in different genes with the repeating segment made up of different numbers of nucleotides, like three (trinucleotide repeats in Huntington's disease and spinocerebellar ataxias including SCA1, 2, 3, 6, 7, 8, 12, and 17), four (tetranucleotide repeats in myotonic dystrophy type 2, or DM2), or five (pentanucleotide repeats in SCA10).

DNA encodes instructions for every biological process. DNA is transcribed into RNA, and then RNA is translated into proteins. RNA and proteins carry out diverse biological processes within cells. You've no doubt seen representations of the lovely double helix form of DNA, and so you know that DNA has two strands that are bonded together like a ladder twisted into a spiral. The two DNA strands are read in opposite directions, and each strand can produce various RNAs.

With respect to any particular gene, like *C9orf72*, the direction that it is read to make a protein is referred to as the "sense" direction. The Ionis therapy is called "antisense" because it consists of a synthetically manufactured sequence of nucleotides (referred to as an "oligonucleotide"—the prefix "oligo" means few), which is designed to bind to the sense strand of RNA that includes the mutation. This double-stranded RNA is then degraded by an enzyme in the nucleus of cells called RNaseH.

However, the Ionis antisense oligonucleotide, or ASO, addresses only half of the problem. This is because in these repeat diseases, the transcription and translation machinery runs amok and creates an unexpected RNA in the antisense direction, as well as the sense direction, of the mutated gene. Each mutated strand of DNA creates two RNAs (and each RNA is translated into proteins), but the Ionis ASO only attacks the sense strand of the RNA and not the renegade antisense strand. This aberrant translation feature of repeat diseases was discovered by Laura and her lab and is called repeat associated non-AUG translation, or RAN translation. More on that later.

Whether the remaining antisense RNA strand will cause problems is one of the key questions with respect to the Ionis therapy. We hope that by reducing half of the troublesome RNA and downstream RAN proteins, we'll see significant improvement. I've had this *C9orf72* mutation all my life, and it only started creating problems as I reached my late 50s. So maybe my normal biological housekeeping processes have been able to clean up these renegade RNAs and protein for a long time until their production increased due to age or stress, and they started to cause a problem. Or maybe my housekeeping biology is not as good at 60 as it was at 25. In either case, if the antisense therapy can reduce the troublesome RNAs by 50%, maybe that will be enough to get me back to a place where the RNAs or the proteins produced by the RNAs do less damage.

We say the mutation "causes" the disease, but we are only beginning to understand how it works. What scientists really know is that the mutation correlates with the disease. They have studied families like mine and found that the people who get sick have the mutation and the people that don't get sick don't have the mutation. Complicating this further, some people who have the mutation don't develop the disease. Identifying a specific mutation that correlates with the disease is a major breakthrough in understanding the disease mechanism, and advancements in gene mapping in recent decades have resulted in more and more disease-related mutations being found. Scientists at the Mayo Clinic in Jacksonville, Florida and the National Institutes of Health in D.C., found the *C9orf72* gene mutation in 2011, about the time that I began to notice changes in my voice. None of this was known when my mom died of ALS in 1989, and I feel incredibly lucky to live at a time when science is advancing toward a cure.

The fight for a cure is fought on multiple fronts, however, and patients who want to be on the front lines have to choose their field of battle. In 2016, I spoke with a doctor who encouraged me to fly to Moscow to receive an injection of stem cells in my spine. The price for this procedure, including hotel reservations, was about \$50,000. Airfare was separate. I first spoke with the doctor in early September 2016, and he and his wife, who served as his assistant, proposed arranging for flights that left two weeks later. The plan was to depart on the 26th and arrive in Moscow on the 27th and then stay overnight at the hotel. I would check into the hospital on the 28th, have the procedure, and stay overnight in the hospital. I would then check out of the hospital on the 29th and spend one more night at the hotel before departing Moscow on the 30th. The email discussing the logistics sounded like an ad for a vacation:

"Most of our patients stay at the Ritz, which is a great location, right across from Red Square. The Ritz has a club level where patients can take all their meals in the private dining room which overlooks Red Square and the Kremlin and is included in the room cost. With the devaluation of the Ruble, the room runs about $400 a night which includes your meals."

I had been referred to the doctor by a client of mine, and so I listened respectfully before reporting that because my passport was expiring within six months, I was not able to make the trip.

I was highly skeptical of the whole thing. I asked the doctor for some information about other patients who had received the treatment and received only anecdotal reports of patients who had quite advanced cases of ALS when they received the treatment and had not survived. There seemed to be no evidence that patients benefited from the treatment. Also, nothing in my discussions with the doctor or the arrangements for the treatment suggested any scientific rigor. There were no papers that evaluated the mechanism of action, safety, or efficacy of the treatment. The treatment seemed to rely on the assumption that stem cells injected into the spinal cord would repair the damaged nerve cells that were causing the problems in ALS. It didn't seem to matter that my disease was tied to the C9 mutation, or that I had bulbar onset. Finally, the $50,000 price tag and the high-pressure sales pitch—the doctor actually said to me "you don't have a lot of time to wait"—turned me off. Laura was aghast that I was even talking to this guy.

I have seen a *60 Minutes* episode about a different doctor who promoted stem cell treatments for ALS patients that turned out to be a scam. Also, in *Gleason*, a wonderful movie about pro football player Steve Gleason's battle with ALS, he gets a stem cell treatment and actually regrets doing so because he believes he got worse after the treatment. We ALS patients are vulnerable because we hope that science will save us. So we are drawn to those who say they can help, some of whom deceive us, some of whom approach the science lazily hoping their treatment will work and also make them some money, and some of whom are great minds driven by curiosity and a desire to help. Luckily, I am married to one of the latter group.

A few days before we left for the screening visit in Baltimore, our good friend Emily Plowman joined us for dinner at our house. Dr. Emily Plowman is an associate professor at UF and studies how neuromuscular diseases like ALS affect speech and swallowing. She was thirty-something, fit, and beautiful, with long, dark brown hair, intelligent eyes, and an Australian accent. I have been visiting Emily's lab and clinic every month or so since March 2016 with a break in 2020 for COVID. She and her colleagues evaluate my swallowing and respiratory strength and capacity. They have the same machine that I saw in the Road Runner and Wily Coyote cartoons—the one they would pull in front of Wily Coyote to provide X-ray evidence of what he had swallowed. In my case, they had me eat or drink various substances including barium and watched as I swallowed the substance. On the screen we could clearly see the substance being chewed and swallowed. The conclusion so far is that I am safe with swallowing. When I swallow something thicker, the consistency of a banana or pudding, I have a little residue that gets hung up in my throat but is washed away with a swallow of water.

Emily and her team also measure my expiratory force and inspiratory force. For the expiratory force, I blow into a device with one explosive breath, and for the inspiratory test, I suck air in through another device. I do each test several times, and they record all the measurements. They also test the volume of my exhaled breath, the "forced vital capacity" (FVC), by having me inhale as deeply as I can and exhale as completely as possible into a device. Then they record me reading certain sentences, to get a measurement of my rate of speech.

Emily has given me two devices to exercise my expiratory and inspiratory muscles. I do each about five days per week, in sets of five, with each set consisting of five breaths. I do the exercises regularly, and I believe I have benefitted from them. Emily and her team have even written a paper about my improvement in various parameters as a result of the exercise. I recommend this exercise program to anyone who would like to increase their volume or force of breath.

Emily knows my situation well. We had invited her to dinner because she had asked to speak to us about the clinical trial. She was concerned. She said, "When I heard about this trial, I just had a bad feeling. I can't really explain it, but I felt I owed it to you guys to let you know. I've seen a lot of patients try something and end up worse off. And you remember the Gleason movie and how the stem cell treatment left him worse off. You've got a good thing going; don't screw it up."

I recalled Malcolm Gladwell's book, *Blink*, and the ability of people with deep expertise in a field to size up a complex situation in the blink of an eye. Emily is one of those people. Emily had also read the book, and we talked about her reaction and what was driving it. I wondered whether we were doing the right thing.

But as we talked further, I regained my confidence that we should move ahead. This treatment is designed for the *C9orf72* mutation that I have. They have preclinical evidence that has convinced the FDA to allow them to move forward into humans. They are starting with a small dose. Jeff Rothstein thinks it's a good idea. The snippets of DNA designed to bind to the expansion RNAs seem unlikely to cause off-target effects.

As we hugged Emily goodbye, we told her that we appreciated her friendship and her honesty in sharing her concerns but that we were going to go ahead and hope for the best.

Chapter 7:

Metformin

Before the screening visit, I had started taking a drug called Metformin. The drug is a widely used diabetes drug that came to Laura's attention at a scientific seminar where it was reported to have an effect on protein translation (the process by which RNA is translated into proteins) in a disease called fragile X syndrome. Knowing that the toxic effect of the *C9orf72* mutation is directly related to abnormal, RAN protein translation, she decided to test metformin on C9 cells, or as she tells it: "We threw it on some cells, and it reduced RAN protein levels!" She and her team also gave it to *C9orf72* mice and they seemed better. This is part of what makes Laura a great scientist. She is constantly absorbing information and integrating it in creative ways into the problems she is wrestling with.

That started the long process of creating a controlled experiment to see if it really works. Before any data from that experiment was available, however, Laura and I started to talk about whether I could get a prescription for metformin from my general doctor. When we talked to the doctor about it and explained our reasons, she was quite surprised and said, "Metformin! They give that away at Publix. That's a safe drug. I could give you a scrip for that."

I've been pretty healthy in the past and have generally been reluctant to take drugs, from recreational drugs to antibiotics to cold medicine. I've been of the view that the human body is a biochemical wonder and that introducing foreign substances is likely to do more harm than good. But now that I have ALS, I know that doing nothing is not likely to end well. In June 2017, I got a prescription for metformin and started taking it twice per day, ramping up the dose to the maximum, 2,000 mg daily, that is prescribed for diabetic patients.

Metformin is derived from the plant *Galega officinalis*, commonly known as French lilac or goat's rue. The plant has a history as an herbal remedy used in Medieval Europe to treat symptoms associated with the disease now known as diabetes. Although the active compound was identified in the late 1800s and the analog named metformin was finally recommended for treatment of diabetes in 1957, the mechanism of action of metformin is still the subject of investigation. Researchers are studying whether metformin can be used for cancer, cardiovascular disease, or even as an anti-aging treatment. The more Laura and I learned about metformin, the more we thought it was probably okay and might do some good.

We spoke to some friends about our experiment with metformin, and one of them sent us an article reporting that diabetic patients in England had a statistically significant lower incidence of ALS than the general population, leading us to speculate that this might be true because most of those patients were likely taking metformin. Preliminary data began to trickle out of the lab continuing to support the theory that metformin reduced RAN translation.

I had no adverse side effects from taking metformin, and with the developing story I was feeling pretty good about taking it before the screening visit. To qualify for the ASO trial, however, I was supposed to stop taking any drugs except Riluzole, a drug approved to treat ALS that I had been taking since my diagnosis. Obviously, to test the new ASO, they wanted to measure its effects without the possible influence of another drug.

Laura and I pondered our choices. We could stop taking metformin despite our growing sense that it was probably a beneficial treatment. We didn't like that. We could continue taking metformin and explain our position to Jeff Rothstein and hope that he could get Biogen to allow us to stay in the trial. We knew that was unlikely. Jeff had done us a great favor by getting me into the trial; I was only one of three ALS patients at Johns Hopkins to be in the trial. There were many others who wanted to be in the trial and would be happy to comply with all the requirements. It was unlikely that Jeff would want to raise the issue with Biogen or that Biogen would agree to such an exception. And asking Jeff would only put him in the awkward position of having to say no to us.

The other choice was to continue taking metformin but say that I had stopped. Laura didn't like this. She said, "I can't do this. I could lose my job over withholding relevant information in a clinical trial."

I understood her position, but playing devil's advocate and in an attempt at sarcastic humor, I said, "Oh, so you're going to take the position of big pharma instead of giving your poor, sick husband a better chance to survive?" It was not the right thing to say. After a moment to wipe away tears and compose herself, she said, "I'm on your side, but this is hard. We have to do this the right way. I think the metformin is helping, but if you have to stop taking it to get in the trial, that's what we have to do."

I wasn't convinced. *What if I get the placebo in the trial?* Then, I imagined, I would simply be another body for Biogen's numbers, letting the disease do its relentless damage, when I could be taking a safe drug that may provide a benefit. *Why would Biogen care?* The trial is designed to test safety, tolerability, and pharmacokinetics, which refers to the absorption, bioavailability, distribution, metabolism, and excretion of the drug over time. If one patient out of 44 were taking metformin, I doubted that would affect the overall results of the trial as to the safety, tolerability, or pharmacokinetics of the ASO they planned to test.

Biogen understands that patients should not be deprived of drugs that may help them and allows patients to continue taking Riluzole while in the trial. If metformin were approved by the FDA to treat ALS, it would probably also be on the approved list. So, I reasoned, I shouldn't have to stop taking it just because it hasn't gone through the proper bureaucratic maze. If an ALS patient had diabetes and were taking metformin for the diabetes, they would probably allow the drug, I speculated.

Finally, I asked myself, *Are you willing to break this rule to give yourself a better chance to survive?* My answer was yes. This presented an ethical issue that I did not take lightly. I thought of all the talks with our kids about the importance of honesty and integrity, and to lie about the metformin would make me a hypocrite. On the other hand, I wanted to do everything possible to fight this disease and doubted that the metformin would have any effect on the overall results of the trial. My plan was to keep taking the metformin and not tell Laura, Jeff, or the Johns Hopkins team.

We returned to the Outpatient clinic at 601 N. Caroline Street on the fifth floor for the screening visit on September 11, 2018. We checked in at the kiosks, waited a bit, and a nurse took my vitals. Then we met with Jeff and were introduced to Lora Clawson, MSN, CRNP, Director of ALS Clinical Services, and Kristen Riley, PhD, Clinical Research Program Coordinator. Lora and Kristen would do most of the work with patients in the trial.

Jeff talked to us about the trial. He explained that it was a Phase I clinical research study and would be the first time the ASO had been tested in humans. There were expected to be 44 subjects in the study and only three at Johns Hopkins. Jeff said that they could have filled the entire trial with just patients from Johns Hopkins, but the drug companies prefer to have multiple centers involved. He explained that the drug had already been tested in animals at doses higher than the dose planned for the trial and that the drug had been well tolerated and there were no serious side effects. The study would be placebo-controlled, meaning that every participant would be randomly assigned to a study drug group and that one group would receive the study drug and the other would receive placebo. In the study, three out of every four participants would receive the drug and one out of every four participants would receive placebo. The study would also be "double blind," meaning that neither we nor the Johns Hopkins team would know whether I was receiving the study drug or placebo.

During this visit, Lora took a blood sample to confirm the presence of the *C9orf72* mutation and for various assessments including measures of how well my blood clots. Kristen placed on the desk a three-ring binder with the largest rings that I had ever seen. It was empty, except for a piece of paper or two. She explained that this was my binder and, yes, they would fill it up by the end of the trial. At some point during the day I went over the medications I was taking: Riluzole, Nuedexta, a supplement called uThrive Agility, a probiotic, and

Vitamin D. Although I felt a little guilty about it, I omitted metformin.

The title of the trial was "A Phase 1 Multiple-Ascending-Dose Study to Assess the Safety, Tolerability, and Pharmacokinetics of BIIB078 Administered Intrathecally to Adults with C9ORF72-Associated Amyotrophic Lateral Sclerosis." BIIB078 is the name of the ASO drug for testing purposes. They'll give it a jazzy, trademarkable name later, when they start asking us to pay big bucks for it. The "Multiple-Ascending-Dose" part of the title refers to the intent to evaluate the safety and tolerability of one dose in a cohort of patients before moving on to a higher dose. I was in the first cohort of patients and so, if I was not in the placebo group, would receive the lowest of the planned range of doses.

Assuming I met all the criteria from the screening visit, they planned to administer the drug five times during a period of 12 weeks. The first drug administration day was Day 1, and there were additional drug administrations on Days 15, 29, 57, and 85. Each drug administration required a lumbar puncture (LP), and during each LP they would also remove some cerebrospinal fluid (CSF) for evaluation. One week after each drug administration, I would be required to return to the study center for follow-up evaluations. After the drug administration visits, I would be required to return to the study center on Day 134 and Day 176 for LP procedures to remove CSF for evaluation. So overall, the study would require 12 trips to Baltimore from Gainesville over about six months, each with an overnight stay, but fortunately Biogen would reimburse us for travel, hotel, and meals.

A week after we got home from the screening visit, however, Kristen emailed us the results of my blood test, and two of the coagulation tests measuring clotting time were above the acceptable range for the trial. She said we could retest only once to get within the acceptable limits. She wondered whether the supplement I was taking might be the reason for the elevated test results, and she asked me about the ingredients for the uThrive Agility supplement. I emailed them to her. This is how the supplement maker describes the ingredients: catechin from green tea, curcumin (potent anti-inflammatory), and sulforaphane (nature's most potent anti-oxidant regulator). uThrive Agility was developed by a professor at the University of Florida, Brent Reynolds, PhD, with whom we've spoken, and we like the local connection.

Laura and I were concerned. If I didn't get those coagulation times down, we could be out of the trial. We could miss an opportunity to slow a relentless and deadly disease. And although Kristen was focused on the uThrive Agility, I wondered if it might be the metformin. A quick internet search confirmed that metformin does indeed play a role in coagulation and seems to decrease the risk of thrombosis in diabetes patients. I concluded, *I need to stop taking the metformin right away so that it will be out of my system by the next blood test.*

Kristen emailed a copy of an article stating that curcumin also had anti-coagulant properties and recommended that I stop taking the supplement for a few days before I got the retest needed for the trial.

Was it the curcumin or the metformin, or both, that was affecting the coagulation parameters? We couldn't take chances, and I stopped taking both. My ethical discomfort was resolved. Five days later I had another blood test. On September 25, 2018, Kristen emailed that my coagulation parameters were in the normal range.

On September 26, 2018, as we were traveling to Minnesota for Paul's wedding, we received Kristen's email that I had been accepted into the trial. Laura and I were relieved and excited. We were anxious to do something to alter the expected course of the disease and believed that this new drug was well designed to do that.

Chapter 8:

Wedding

The wedding was on September 29, and my first injection for the clinical trial was on October 2, and in the months leading up to the wedding, I worried about the condition of my voice. My brother, Mike, gave wonderful, heartfelt talks at events in connection with his two daughters' weddings, and my brother-in-law, Clark, gave a similarly touching speech at the reception following his son's wedding, and so I wanted to say something to honor my son and his bride at their wedding. But when Paul and Susan announced their engagement on July 29, 2017, and began planning for a wedding in the fall of 2018, I wondered whether I would still be able to speak at the time of the wedding.

Fortunately, my voice stayed about the same—a speaking rate of about half of normal speed, but still largely intelligible—and weeks before the wedding, I started to think about what to say and jotted notes down. On the day of the wedding, my daughter Maddie and I practiced our speeches to each other in the small bedroom next to the ironing board, secluded from the general chaos of the wedding party as they prepared in our house (with only a few interruptions from people wanting to iron something).

The wedding unfolded like a fairy tale. The bride was beautiful and the groom handsome. They said their vows at St. Olaf Catholic Church in downtown Minneapolis and, in lieu of departing in a fancy car after the ceremony, took a photo op ride on a tandem bike. The reception followed at the Walker Art Center near downtown, and dinner was served to the 186 guests in an elegant room with windows overlooking the lights of downtown.

I ate my dinner with nervous anticipation, telling myself to just do the best I could and be satisfied with that. The first speaker after dinner was the father of the bride, Bill, who is a professional actor and runs a theater. In a wonderful, rich, baritone voice, he told charming stories of Susan as a girl, and as he spoke, a lead weight developed in the pit of my stomach at the prospect of following such a talented speaker. When he finished, the wedding organizer beckoned to me, and I rose from my chair and walked to the microphone.

As I started to speak, I realized that the crowd was chatting and laughing and generally having such a great time that I couldn't help but smile and speak more loudly to get their attention. I said, "My name is Bert Ranum, and I am Paul's dad," and to my surprise, they clapped and cheered. I started to think this might be fun. "I have good news and bad news," I went on. "The good news is I have a short speech. The bad news is it will take me a long time to say it.

"Thanks to all of you for joining us today to celebrate the marriage of Paul and Susan. We are so delighted to have Susan join our family." I turned to look at Susan and Paul, who were sitting slightly behind and to the side of the microphone.

"We have enjoyed getting to know Susan over the past few years while Paul and Susan have been together, and it's clear that he has chosen a strong woman to share his life." The crowd laughed and clapped.

"And in this, Paul is following a tradition: The Ranum men marry strong women. By 'strong,' I don't just mean physically strong. I'm talking about life force, making a positive difference in the lives of others, independence and integrity. Yes, the Ranum men marry strong women."

Here, because I was consulting my notes only infrequently, I left out my best joke, which I can now add in the retelling: *In the Ranum family, the men make all the big decisions, and the women decide everything else, including which decisions are the big decisions.*

I said, "It started with my mom. I've been thinking about her a lot recently because she died of ALS at 62 and now I have ALS.

"She was an intelligent, compassionate, hard-working woman who could be a loving disciplinarian to kids and pets. She rarely raised her voice, but both kids and pets knew she was in charge.

"It was clear to us that the relationship between Mom and Dad was an equal partnership, and she was respected and loved by everyone in the family. She was a strong woman.

"It was natural, then, for my brother Mike and I to be attracted to women with intelligence, energy, and ambition. For those of you who know Mary and Laura, you know that we got more than we deserved.

"Mike's wife Mary is an accomplished lawyer and until recently was the chairperson of the Board of Directors of Fredrikson & Byron, a major Minnesota law firm, where I have been fortunate to be her law partner.

"And Laura," I paused for emphasis, "Laura is known internationally as one of the leading researchers in neurodegenerative diseases." And here I ad-libbed "including ALS." I continued, "She's going to find a cure any day now. She is the director of the Center for NeuroGenetics at the University of Florida.

"On top of these demanding careers, they have each raised wonderful kids and been the center of their families. As I said, the Ranum men marry strong women.

"Which brings us to Paul and Susan. Well done, Paul, at keeping the tradition alive.

"Once again a Ranum man has married a strong woman. Susan is extraordinary, even in a family of strong women. She was raised on grit and tenacity by leading the life of a gymnast much of her youth, and then continued to excel in athletics and academics at St. Olaf.

"I have to point out a particular accomplishment because Mike and I both competed in track and field all four years at our colleges and dreamed of MIAC (Minnesota Intercollegiate Athletic Conference) championships but never won one. Susan was the MIAC champ in the triple jump her sophomore year at St. Olaf! Finally, we have a conference champion in the family!" The crowd laughed and applauded.

"And now she is excelling as a medical student! But even more importantly, Susan lights up a room with her smile and is completely charming.

"So, in closing, let me offer some advice to these two extraordinary individuals who have now become one family."

Here, I turned to Paul and Susan and said, "Be kind to each other—so often we come home at the end of the day tired, and we forget to be kind to those we love the most.

"Be generous to each other—generous in spirit and generous with your time.

"And finally, be quick to forgive each other.

"If you do that, 35 years from now you may be able to say to each other, as I can to my wife…" I turned to look at Laura. "…that I love you now more than ever.

"So now please join me in raising a glass to Paul and Susan, and their future together."

Based on the applause, I thought I had done okay. I turned and found tears in the eyes of both Paul and Susan as they hugged me. I returned to my seat at the table, and Clark shook my hand and said, "You hit the ball out of the park. Great speech!" Mike and Mary came up and hugged me, saying I had done a wonderful job. Laura reached across the table with teary eyes and clasped my hands. "Great job!"

I know that the reaction to my speech was partly driven by sympathy (I had played the ALS card brazenly), but I accept every ounce of good feeling that comes my way. It lightens the burden. We are social animals, and we all need the support of others. I am thankful that I have a wonderful family that loves me.

Maddie spoke a bit later and was beautiful, funny, engaging, and full of love for Paul and Susan. The crowd adored her and applauded and cheered loudly.

After her talk, I could relax and enjoy the rest of the evening, including the band and dancing that followed.

The post-wedding brunch the morning after the wedding was the last Ranum family event at 2116 Carter Avenue in St. Paul. We had bought the house in 1988, the year that Paul was born, so it was the only home our kids had known. As we left the house the next morning headed for Baltimore and Johns Hopkins, we felt as though the earth was spinning a little faster and that change was upon us.

Chapter 9:

First Injection

By the time of the first injection on October 2, 2018, we had been to Baltimore several times and had made the Inn at Henderson's Wharf our regular place to stay. The Inn is located in the Fells Point neighborhood of Baltimore, which dates back to 1763. The cobblestone streets are lined with historic buildings now filled with restaurants, bars, and shops. It started as a shipbuilding center on the harbor, and during a few strolls around the area, we learned from historical signs that a young Frederick Douglass had worked in the shipyards as a caulker, caulking the seams between the wooden planks of hulls to make them watertight. The job allowed him to build on the meager learning he had been able to cobble together by learning the letters that carpenters marked on the timbers; e.g., SF for starboard forward, or SA for starboard aft.

Although Baltimore may have its troubles, based on what I read in the news, it treated us well. The Fells Point area is charming, and Johns Hopkins is a world-class medical center. We also enjoyed the Thames Street Oyster House, a great restaurant located in a 19th century building not more than 20 feet wide, within walking distance of our hotel. While reservations were recommended if you want a dining table, we liked to take advantage of the high-top tables near the bar, which were available on a walk-in, first-come, first-served basis. We enjoyed many meals there in a convivial atmosphere with only a narrow walkway traversed frequently by friendly waitresses separating us from patrons seated at the bar. After COVID, the restaurant put up a canopy in the street and we took advantage of the outdoor dining.

We arrived at the Outpatient Center, fifth floor, shortly before 8:00 a.m. on October 2 for our appointment, checked in at the kiosks, and were met by Kristen. We had been told that the first injection would take place in the hospital and that I would be required to stay overnight, and consistent with this, Kristen walked with us to the hospital and checked us into a room. Lora appeared and took my blood pressure and pulse, obtained a blood sample, and gave me a cup for a urine sample.

At about 10:00 a.m., Dr. Michael Levy joined us. Dr. Levy was a good-looking, 40ish man with trim, dark hair and an overall air of calm, friendly competence. I imagined that he had done a lot of LPs, and this gave me confidence. He had me sit on the edge of the bed and bend over. The injection of the anesthetic, lidocaine, stung a bit, but the actual LP was painless, and I could only feel a little pressure, which let me know there was something taking place in my lower back. Before administering the dose of the ASO (or placebo, as the case may be) of 5 milligrams, the study protocol called for them to withdraw 10 milliliters of CSF for evaluation. The procedure was over in about 30 minutes.

After the procedure, Laura and Dr. Levy had a short discussion about post-LP patient care. Dr. Levy suggested lying flat for a couple of hours. Laura had heard or read somewhere that getting up and walking immediately after the LP was helpful to avoid post-LP headache, which is a common occurrence. Dr. Levy said he was aware of no studies that indicated a benefit to either approach, but advised taking it slow, if we chose to walk. So shortly after the procedure, I got up, and Laura and I took a slow walk around the hospital floor that we were on. Then I returned to the room and lay down.

Lora and Kristen had several post-LP duties to be carried out on a schedule. Lora took blood from me at one, two, four, and six hours after the dose and checked my pulse and blood pressure at two and four hours post-dosing. She also did neurological exams at three and six hours post-dose. Kristen took an electrocardiogram (ECG) at two and four hours after the dose. I felt completely normal. All of this was carefully recorded.

After the sixth-hour activities, Laura and I were hungry, and on reviewing our food options with a nurse, we learned that in addition to what the hospital offered, we could also get a takeout order from a Thai restaurant nearby. Laura went out for the Thai food, and we enjoyed our pad thai with chicken and egg rolls there in the hospital room. Then Laura left to go back to the hotel, and I spent the night in the hospital.

In the morning, at 24 hours after the dose, Lora and Kristen asked me how I was doing and again took a blood sample, urine sample, blood pressure, pulse, and ECG and conducted a neurological exam. Everything was good, and we were released from the hospital.

The headache started after we had gotten home to Gainesville. It felt like my head was fragile. Maybe I only imagined it because I knew that CSF had been withdrawn from my spine, but I felt as though my brain did not have sufficient fluid around it to cushion sudden movements. I tried to be still. Sneezing or coughing were particularly painful. Thankfully, the pain was relieved by lying down, so sleeping was not a problem.

Laura is my hero and my best advocate, but I was in pain and not above blaming her for her recommendation to walk immediately after the LP. "Next time, we are definitely not going to walk," I said.

The headache lasted about two days, but it was bad enough that I was thinking this trial was not going to be a picnic if this happened after every injection. Laura turned to the internet, where she learned that the headache is said to result from the leakage of CSF from the puncture in the dura (the fibrous sheath surrounding the spinal cord) left by the needle. Conventional LP needles have a tip that is shaped like a straw cut at about a 20% angle to its long axis (designed to cut through tissues of the dura), while atraumatic needles have a cone tip like a pencil point with a port on the side of the needle for fluids to pass. Atraumatic needles more often separate and dilate surrounding dural fibers rather than cutting through them. Subsequent contracture of the fibers after needle removal results in a small pinpoint opening in the dura, as opposed to the irregular and larger opening created by conventional needles.

For the rest of the LPs we asked for, and got, atraumatic needles. And the headaches went away.

About a week after the first dose, we received an email from Kristen reporting that my blood work showed high coagulation times again. Since I had stopped taking the uThrive supplement and the metformin some time ago, we wondered if it could be the Thai food that we had eaten in the hospital that caused the abnormal results. We suspected that the Thai food might have had monosodium glutamate (MSG) in it, and we did some internet searching on whether MSG could affect blood coagulation. We were surprised to find that it could, and reported this to Kristen. Kristen emailed back that even the bloodwork before the dose (and so before we had eaten the Thai food) had the abnormal readings, so it couldn't be that. This mystery resurrected the worry that they might take me out of the trial.

At about the same time, a week or so after the dose, I began to realize that my peeing urgency had diminished significantly. (If this is too much information, skip the rest of this chapter.) For at least the last year, I had been noticing an increasing level of urgency when my bladder was full, or even when it was not full. I could be perfectly comfortable one moment, then suddenly I would be seized by a need to pee so aggressively that it would make me grimace, and I would begin to panic about whether I could find a bathroom. I began to assume that my window of safety was a max of two hours. Long car rides and airplane rides were especially problematic. I learned to look for prophylactic opportunities to stop at restrooms when they were available.

I had assumed that my peeing problems were just part of aging and possibly due to an enlarged prostate. My internist said my prostate was about typical for a man my age and gave me a referral to a urologist. The urologist said the same thing and gave me samples of a drug called Myrbetriq for overactive bladder. I never took the Myrbetriq because we got word about the trial shortly after my visit to the urologist.

The change in my urgency symptoms was so striking that even though I had never associated the problem with ALS, I concluded that *the improvement must be due to the medication in the trial, and that meant I was not on placebo!* I have since learned that ALS can cause an overactive bladder. As we all know, the bladder is full of nerves, and if there is irritation or damage to those nerves, a variety of urinary tract symptoms can result, including an overactive bladder. So this change in my symptoms convinced me that I was on the drug, and even more importantly, that it was working to decrease nerve damage. I imagined that there were even more benefits that were beyond my perception.

At my next appointment, I explained my observation to Lora. She confirmed that ALS sometimes caused bladder issues and seemed mildly interested in my explanation of the change in my symptoms. But it did not seem to be as significant to her as it was to me, nor did it seem to be recorded anywhere. I was a little surprised that my experience, which was a material improvement in my quality of life, wouldn't be of interest to the people developing the drug, even though it was anecdotal. Plenty of the information they gathered was subjective patient assessments. The ALS Functional Rating Scale Revised (ALSFRS-R), which is the widely accepted gold standard for assessing the progression of ALS, asks the patient to rate various measures like speech, swallowing, and handwriting on a scale of 0 to 4, where 0 is complete loss of function, e.g., 0 is loss of useful speech and 4 is normal speech. I think they should add a "urination" measure to the ALSFRS-R and ask patients to assess their peeing function on a 4-point scale.

At that second appointment we also learned the answer to the blood coagulation mystery. Because another patient's blood results showed high coagulation times, Lora began to suspect an error in the lab tests. They had the samples retested at another lab, and they were normal, so the first lab had somehow made an error. Like all complex endeavors, there are a lot of things that can go wrong in a clinical trial.

I am lucky to be in this trial at Johns Hopkins. Johns Hopkins has the history, reputation, facilities, and most importantly the team of excellent doctors, nurses, clinicians, and administrators necessary to earn itself a place in the big pharma trials. This doesn't happen overnight. Organizations evolve over time, and the best begin with a culture of striving for excellence that holds everyone to a higher standard. Because everyone, or at least the leaders of the organization, are striving to be better, over time the organization becomes better.

Although Dr. Jeff Rothstein is the leader of the team at Johns Hopkins involved with the trial, Lora Clawson and Kristen Riley were the key people who collected the data and with whom I interacted at each visit, and they were excellent. Lora Clawson, MSN, CRNP, is the Director of ALS Clinical Services. Lora has worked with Jeff for over 30 years and combines personal warmth with brisk professionalism. She has two golden retrievers that she talks about in a way that caused me at first to think they were her children. In my defense, they are named "Charlie" and "Sally," which adds to the likelihood of confusion.

Kristen Riley, PhD, is Clinical Research Program Coordinator. Kristen has the informality and confidence of a scientist trained in the crucible of the academic world. On the inside of her left wrist, she has an inconspicuous tattoo of a lightning bolt and a pair of glasses marking her as a devoted Harry Potter fan. Kristen is all business, often juggling several things at once, such as attaching leads for an electrocardiogram (ECG) while also taking a cell phone call about something else.

I saw Lora and Kristen on every visit. On several visits I also saw Alpa Uchil, a registered nurse who is part of the team. Alpa is very kind. For example, she was concerned about the amount of discomfort I suffered from pulling off the adhesive patches for the ECG together with a bit of my chest hair. I told her that I have been through worse, but I appreciated her sympathy nevertheless. In 2018, Alpa was taking classes after work in the evening to become a nurse practitioner, and she had completed her studies by the time of my later visits and graduated to giving me lumbar punctures from time to time. I admire her ambition and her dedication to nursing.

Apparently, Johns Hopkins has rules about who can perform the LP. During the trial and the later open label extension, most of my LPs were performed by Arita McCoy, MSN, CRNP. Arita is calm, which is what you want in a person who's going to stick a needle into your spine. I got the impression that she had done this a lot, and she was very relaxed and soothing in her manner, which put me at ease. They asked me whether I wanted to be on the examining table for the procedure, and I said, "I want to be wherever Arita wants me." She preferred that I sit in a massage chair. I've seen people in these chairs getting a massage in malls and airports that offer massages, but my first experience in a massage chair was for an LP. Whether in a massage chair or on the table, the idea is to raise the knees and curve the spine in a fetal-like position, to create more space between the lumbar vertebrae.

Susan, my new daughter-in-law who is a medical school student, tells me that lumbar punctures are a rite of passage for medical students. The fluid that circulates in the brain and spinal column, cerebrospinal fluid, or CSF, is clear and colorless, and med students call a first LP that produces only clear fluid with no blood a "champagne tap." The event is often celebrated with a bottle of champagne.

There was no champagne involved for any of my LPs, but Laura told me that the CSF that dripped from the needle was always clear. In the LP, a hollow needle is inserted between the vertebrae in the lower back, in my case between L3 and L4, and into the space at the bottom of the spinal cord called the "cauda equina," which is Latin for horse's tail, apparently because that's what the nerves in the area resemble. There is no risk of damage to the spinal cord itself because the point of insertion is below the spinal cord. For my LPs, they always collected some CSF for testing, and on those days when I was receiving a dose of medication, they injected the medication through the same needle.

The procedure sounds worse than it is. The most uncomfortable part was the injection of the anesthetic, lidocaine, in preparation for the LP. During the LP itself, I felt only a little pressure and knew what was going on only by what others told me. Arita might say, for example, "Okay, I'm in. Are you doing okay?" Then I just waited, remaining very still. The collection of CSF took a few minutes because they relied on the pressure in the spinal column to cause the CSF to drip out of the needle where they collected it. It is just like collecting maple syrup from a tree. The whole procedure takes about 30-45 minutes.

Lora and Kristen are constantly working on clinical trials for ALS treatments. In addition to my trial, they had another trial going for another Ionis/Biogen ASO for ALS patients with the SOD1 mutation. They work on about 25-30 trials per year. The big pharmaceutical companies know from experience that Lora and her team can be relied upon to follow protocols and produce reliable data consistently. They are a big reason that Johns Hopkins is typically selected as a participant in the trials.

Lora, Kristen, Alpa, and Arita—I love these women. They are the unsung heroes fighting in the trenches to help people like me. They balance their intelligence and energy with empathy and compassion. They don't make big money, but they make a big difference. And they work hard. I hear it in comments from time to time, not complaining, but hints of weariness from the grind of dealing with the schedules, appointments, deadlines, equipment, night school, kids, dogs, husbands, and troublemakers like me. How do they carry on without becoming cynical? We ALS patients don't have a lot of happy endings.

The science must fascinate many of them. It fascinates me. Each trial is a biological experiment. Patients' lives hang in the balance. There is some drama. But I suspect the drama is lost in the day-to-day slog of seeing patients and recording the minutiae required by the study. The Johns Hopkins team, except for maybe Jeff Rothstein, doesn't get information on how the study is going. It's a one-way street: Johns Hopkins reports on their patients to Biogen, and Biogen does not give summaries of what they're seeing overall to Johns Hopkins. So Lora, Kristen, Alpa, and Arita have to be content with gathering data and waiting months for the result.

They must also feel good about helping others. Doing something interesting in the service of others is really the best we can do with our lives, isn't it?

Chapter 10:

Publications

The morning of my fourth injection, I woke and found Laura at her laptop in a corner of the hotel room. When she saw that I was awake, she said excitedly, "I just got an email from Stella Hurtely, an editor at *Science*, asking me to review a paper about double-stranded RNAs triggering a pathway…" This was incomprehensible to my still sleepy brain, which even wide awake is not fluent in the language of science. I asked her to slow down, and after she repeated it a few times I understood the basic outline. To understand the reason Laura was excited about Stella's email, I have to take you on a bit of a digression, so bear with me.

Laura and her team at the lab had submitted what they considered a groundbreaking story to *Science* magazine about a study they had been conducting in collaboration with Biogen, the giant pharmaceutical company, and Neurimmune, a small, Swiss biotech company focusing on antibodies from people that Neurimmune calls "superagers," those unusual people who are in good health in their 90s and even over 100. Neurimmune had provided several antibodies to Laura's lab, and the lab evaluated their efficacy in attacking the RAN proteins produced by the *C9orf72* mutation. Laura's lab did this by injecting the antibodies into mice whose DNA had been modified to have the *C9orf72* mutation from my family. I think of the mice as cousins, noble soldiers in the war on ALS, and I respect their sacrifice. It was no small feat for Laura's lab to produce these mice, which get largely the same symptoms as humans with ALS, progressive muscle weakness and cognitive decline with eventual paralysis and death. So far, they are the best mouse model of *C9orf72* ALS available and an important contribution to finding a treatment.

In short, the study concluded that a particular antibody to the Guanine-Alanine dipeptide protein, one of the toxic proteins produced by the *C9orf72* mutation and known as the "GA protein," when administered peripherally (under the skin), finds its way into the brain and attacks the GA protein aggregates there. More importantly, by several measures, the mice do better and survive longer. This is big stuff, especially for those of us with *C9* ALS.

Laura and her lab drafted a paper with all the data and submitted it to *Science* on May 24, 2018. Science responded with comments from three reviewers on June 27, 2018. Reviewers one and two were fairly positive and had few comments. Reviewer three was more skeptical and had a lot of comments that required additional work to produce more data in response. Reviewer three had even suggested doing a human clinical trial, which obviously is way beyond the capability and budget of an academic research lab. On the morning of my fourth injection, November 28, 2018,

Laura's lab was putting the finishing touches on the response to the comments and expected to resubmit it to *Science* soon.

At the same time as the Biogen/Neurimmune study, Laura's lab had been working on another therapeutic approach using metformin, which as I explained earlier I had begun taking but then stopped when the trial began. To see if it really worked, the lab turned to my little cousin soldiers, and fed some metformin and others the regular diet. To their surprise, the metformin-fed mice seemed to do better. They also had reduced motor neuron loss. A paper was drafted with the results of the metformin experiment and submitted to *Nature* on June 19, 2018. Surprisingly, and unfairly, it was sent to only one reviewer, and that reviewer rejected it.

When Laura and her lab decided where to resubmit the paper, they concluded that they should submit it to *Science* because it supported the antibody story. Both the antibody and metformin result in fewer toxic proteins circulating in the body and an improvement in the mice. The antibody directly attacks the proteins themselves, and the metformin works further upstream by reducing the amount of proteins that are produced. Both point to the toxic RAN proteins as the problem.

When Stella Hurtley emailed Laura that Wednesday morning, Laura was planning to submit the antibody paper and the metformin paper to *Science*. By coincidence, the paper that Stella asked Laura to review also related to protein translation. Laura was excited because Stella's email gave her an opportunity to respond and give a little advertisement of coming attractions; i.e., tell Stella that she expected to resubmit the antibody paper shortly and that she would also be submitting a complementary paper on protein translation and metformin. Laura emailed back that she would be happy to review the paper that Stella had asked her to review, but also said she would understand if Stella concluded that Laura had a conflict of interest given that she was also submitting a paper dealing with protein translation. Stella quickly responded that she would put down Laura as declined on the review and that she looked forward to seeing the two papers. Laura resubmitted the antibody paper to *Science* on December 22, 2018, and submitted the metformin paper for the first time to *Science* on January 5, 2019.

On January 31, 2019, Laura received notice that *Science* rejected the antibody story. The revision submitted earlier had largely satisfied reviewers one and two, but reviewer three had dug in and provided eight pages of additional comments. Laura was surprised at the aggressiveness of reviewer three's comments. He or she seemed skeptical of the data and mentioned a notice on the Jackson Laboratory website that my mouse cousins were not showing the symptoms that were expected.

Jackson Laboratory in Bar Harbor, Maine, collects strains of research mice from all over the world and makes them available to other researchers. The academic community is based on exchange of information, and Jackson Laboratory fulfills a central role by maintaining and distributing mice developed for research purposes. Laura's lab sent my mouse cousins to Jackson Laboratory to make them available to other researchers, but was surprised to learn that Jackson Laboratory had criticized the model publicly, and even more surprised that reviewer three thought a mention on a website was relevant to the antibody paper. Although the Jackson Laboratory mice were derived from mice from Laura's lab and should have the *C9* mutation, the website provided no data, just a sort of web-based "tweet" that raised questions about the mice.

Laura estimated that reviewer three had spent a considerable amount of time and effort building a negative case against the paper, raising questions in Laura's mind as to reviewer three's objectivity. Could he or she be a competitor who couldn't bear to see Laura's work published in *Science*? Laura has learned that the ivory towers of academia are not free from petty rivalries. Collaborators at Neurimmune and Biogen also puzzled over reviewer three's comments.

On a scientific paper the first and last positions in the list of authors are the most important. The last name is always the principal investigator, the lead scientist in whose lab the work is conducted. Here, that was Laura. The first name in the list is generally the scientist making a substantial intellectual contribution whose hands-on work drives the project. Other contributors to the work are listed in the middle. The first author on the antibody paper submitted to *Science* is Lien Nguyen.

Laura says Lien is one of the best trainees she's ever had. Lien grew up and went to college in Vietnam and then did her graduate work at the University of Illinois before coming to Laura's lab at UF as a post-doc. She stands about five feet tall and has a warm, gentle smile that belies the energy and ambition driving her indefatigable efforts in the lab. Everyone in the lab knows I have ALS and is gracious to me, but Lien is particularly solicitous of my welfare. She always greets me with a hug and asks how I am doing. It is always a pleasure to see her. Her work may save my life.

Lien devoted several years of work to that paper, and she was deflated when *Science* rejected it. Laura was disappointed also, but this is not her first rodeo, and she has been toughened by lots of rejection in her career. Laura and Lien are pros, however, and the next day they were strategizing about where to submit the paper next.

Science has an affiliate called *Science Translational Medicine* that is also a well-regarded journal. Laura and Lien transferred the paper to *Science Translational Medicine* on February 7, 2019. Laura chatted with an editor at *Science Translational Medicine* at a conference in Lisbon, Portugal in March 2019, where Laura presented the data from the antibody paper, and although the editor was encouraging about their story, their hopes were dashed again months later.

When they received comments on May 17, 2019, it appeared that the paper had again been reviewed by the now infamous reviewer three, and *Science Translational Medicine* rejected the paper.

Lien reorganized the paper, and she and Laura addressed every reasonable comment they could. Then they submitted to *Neuron*, a well-regarded journal, on July 6, 2019. Lien and Laura were pleased when they received comments on the antibody paper from the *Neuron* reviewers. The comments were largely positive and proposed reasonable suggestions. They revised the paper to address the comments and resubmitted it on November 1, 2019. Neuron published the paper on December 9, 2019.

As for me, the important question is not where or whether the antibody paper is published, but whether Biogen and Neurimmune move forward to develop the drug. So far, they seem to remain committed to the project. After all, the data is the data. The mice get better with the treatment.

The metformin paper received better reviews from the *Science* reviewers than the antibody paper. Reviewer three from the antibody paper did not appear to be involved in reviewing the metformin paper. One of the reviewers suggested an experiment that required Laura's lab to develop a new technique, which took several months. When the paper was resubmitted, however, *Science* rejected it.

Laura and her team decided to resubmit the metformin paper to *Proceedings of the National Academy of Science* (*PNAS*), a respected journal but a notch below *Science* on the prestige scale. One of the authors on the paper is a member of the National Academy of Science, which made the submission easier. That author also suggested that he would find someone to write a commentary on the article. The paper appeared in the online version of *PNAS* on July 20, 2020, with commentary by Michael Rossbach, winner of a Nobel Prize in 2017. Laura was thrilled that Rossbach was willing to write the commentary and considered that even better than if the paper had been published in *Science* without the commentary.

The first author on the metformin paper is Tao Zu, a research assistant professor who works in Laura's lab. Tao and his wife Weihong are about our ages and like family to Laura and me. Tao is tall and thin, with a dignified but friendly reserve. Weihong, also a scientist, is much shorter and more expressive, ready to engage or laugh if given the opportunity. Tao first started working with Laura in 2005 when she was still at the University of Minnesota, and in 2010 Laura was thrilled when Tao and Weihong agreed to move to Florida so that Tao could continue to work with Laura. Tao earned his M.D. degree in China and worked for a time in Italy and California before moving to Minnesota and then Florida. Tao was the first author on the 2011 breakthrough paper that discovered RAN translation, and Laura feels incredibly grateful to continue to work with Tao because he is an outstanding scientist and a tremendous asset to the lab. Tao is referred to in Laura's lab as "The Wizard" and is generous with his time and the advice he provides in response to the frequent requests for help on experiments that aren't working. "Ask Tao" is a common suggestion. Tao's work may save my life.

Those of us who are not scientists often assume that science is an objective world where clear-eyed, intelligent people weigh the evidence for and against theories and seek the truth regardless of the consequences. Sadly, that is not the case. According to 19th century German philosopher Arthur Schopenhauer, "All truth passes through three stages: First, it is ridiculed. Second, it is violently opposed. Third, it is accepted as self-evident." By that standard, Laura should have been pleased when *Science* rejected the antibody and the metformin story. It means she is really on to something.

Chapter 11:

The Grim Reaper

I saw a cartoon in *The New Yorker* recently in which the Grim Reaper, with black hood and scythe, stands next to a man who is reaching to squirt some hand sanitizer in the palm of his hand. The Grim Reaper says, "Don't bother."

When I was healthy, I thought about death occasionally and often in the context of the Lutheran/Episcopal worldview in which I was brought up and spent a good portion of my adult life. But like most of us under the age of 60 who are healthy, I was secure in the knowledge that I had a long life ahead. Every death in my family, even my mom's at age 62 from ALS, was very sad, but did not shake my fundamental sense that I was okay. I think we have evolved to discount remote risks. The early humans who were able to stay calm enough to leave the cave and gather food even though they knew the tigers were out there somewhere, were the ones who survived. Now, in the modern world, we don't care about global warming because it hasn't presented itself as an immediate risk to enough people. Although I knew intellectually that my life would one day come to an end and even that there was a chance that I would develop ALS, it didn't really change my behavior.

Since I began developing symptoms and particularly after my diagnosis, however, my perspective has changed. I look at everything with the knowledge that my life may be coming to an end. This change in perspective is not a bad thing. Like driving a car on a beautiful mountain road with a steep cliff at the side, it heightens the senses. I am more grateful for everything every day.

Now those around me are also concerned about my death. My son sent me a wonderful book, Atul Gawande's *Being Mortal*, which I recommend to everyone. It deals with the question of how far we should go to sustain life when medical intervention severely diminishes quality of life. Regarding spiritual matters, my dear sisters, who are fervent Baptists, are trying to save me.

We grew up in a family that went to Trinity Lutheran Church in Crookston, Minnesota, made of limestone (the same limestone many of the buildings at St. Olaf are made of) and one of the prettiest buildings in Crookston, but not every Sunday. The pastor's son, Nathan Hansen, was on the state champion mile relay team with my brother Mike and me and was a good friend. We got the usual dose of Sunday school and enjoyed the church holidays. But after my oldest sister, Calien, was married at 19, she and her husband moved to South Dakota and joined a Baptist church. Later she converted my other sister, Vickie, to the Baptist church, and they have stayed rooted in the Baptist church since then.

The Baptists are very clear about death. Death is not the end. You can beat death and have everlasting life with God if you meet certain criteria, or you continue your afterlife in hell without God. The qualifying criteria are that you must acknowledge that you are a sinner and accept that Jesus Christ died on the cross as a sacrifice for your sins.

To be clear, I have done all that. When I was about 15 and the family visited Cal in South Dakota, she brought us to her church. I don't remember much until the pastor started the altar call. This was startling. He was actually asking us to do something individually. He was breaking the code that had been observed in the Lutheran church, which was that sheep were allowed to stay in the flock, anonymous. He said, "If you want a more personal relationship with God, come forward. If you want to know Christ as your savior, come forward. If you want to feel the joy and freedom that comes with forgiveness of your sins, come forward." I thought, *Well, sure, I want all that, but I'm staying right where I am.* Then my brother, Mike, left the pew and started up the aisle to the altar. My mind raced: *Mike's going?!* My sister elbowed me and whispered, "Go ahead" with a smile. There was no safety in the flock, either. I had to go. I went up to the altar and kneeled, and the rest was anticlimactic. The pastor asked us if we accepted Jesus

Christ as our savior, and we said yes, and he pronounced us saved.

So, I've been there, done that. But my sisters apparently don't believe that the first time stuck, or maybe they've forgotten about that incident. Vickie has recently sent me pamphlets from her church and made me promise I would read them (and I did). Cal wrote a note in her holiday card this year urging me to seek a closer relationship with Jesus Christ as my savior. Fortunately for me, their evangelistic efforts are moderated by their competing instincts to not confront personal issues openly, not give offense, and general inclination toward reticence, which was how our family operated when we were growing up.

When I think about death, I think more about practical things in this life than the life after death. How will my quality of life be at the end? Will I have to get a feeding tube? Will I have to get a ventilator with a tracheostomy? How far should I go in sustaining my life after I lose the ability to walk or talk or communicate fluently?

I understand and share the desire to believe that death is not the end. Our conscious awareness of the world seems separate from the body that sustains us. We dream and experience people and places that are not part of this world. It seems natural to perceive ourselves as souls that may inhabit a body only temporarily and then continue once that body has died. Because we have evolved to be curious and social, we have tried to reconcile this sense of immortality to the world around us by weaving elaborate stories out of the world we knew and the world we hoped for, to explain what we could not understand. These became religions. Every human society on earth has done this.

At the center of most religions is that inherent desire to, and expectation that we will, continue in some way after death. The Buddhists and Hindus believe that after we die we are reincarnated into a new body and a new life in an ongoing cycle of death and rebirth. The Muslims also believe in an afterlife, and like the Christians, they believe the quality of the afterlife depends on how you behave in the current life.

Unfortunately, some people have always used religion and the desire to achieve an afterlife as levers to gain social status, wealth, and power. Throughout history people have been fighting over religion: Catholics fight Protestants, Christians fight Muslims, Shiite Muslims fight Sunni Muslims, Muslims fight Hindus. Each group believes that God is on their side and that by engaging in the righteous battle, they will gain eternal life and probably also status and wealth. More recently, religious fanatics convince the vulnerable that they'll achieve a glorious afterlife if they strap explosives on their body and blow themselves up in a crowd of unbelievers.

Religion and ideas about the afterlife have been involved in so many conflicts, killings, and wars that it's tempting to imagine something different. John Lennon's song "Imagine" has become a secular anthem for peace played in Times Square in New York every New Year's Eve. In it, he sings:

> Imagine there's no heaven
> It's easy if you try
> No hell below us
> Above us, only sky
> Imagine all the people, living for today
> Imagine there's no countries
> It isn't hard to do
> Nothing to kill or die for
> And no religion too…

But if religion weren't available as a lever, the unscrupulous would use nationalism, racism, economics, or sports to divide us and try to advance in the world. The current state of U.S. politics bears this out.

In my opinion, religion says less about the world around us than it says about us. Our religions all say there is an afterlife because we believe that there should be. But in my 63 years on this earth, I have learned that my intuition about something is not always correct; getting to the truth often requires more work. When I look at the moon on a clear and quiet night, some part of me finds it hard to believe that men really stood on that glowing orb in the sky. But I have overcome that initial incredulity with the weight of the information that I have seen and read about that amazing moon landing about 50 years ago, all of which is consistent with my understanding, basic though it may be, of the earth, its atmosphere, its place in the solar system, and the moon that orbits the earth; consistent also with my understanding of the nature of technological advancement from the Wright brothers at Kitty Hawk to Lindbergh crossing the Atlantic, to airlines traveling regularly across the globe and rockets launching satellites into space to landing a man on the moon. Because of all of that, the web of consistent facts that I know, I can add the moon landing as an additional piece of the puzzle and find that it fits neatly in my view of the world.

The afterlife, however, is a piece of the puzzle that does not fit easily in my view of the world. It does, however, fit neatly into what I know about humans. Our religions provide for an afterlife, because most of us don't want to die.

Although I am skeptical of religious doctrine, the more scholarly approach to investigating the question of the afterlife reflected in the book *After* by Dr. Bruce Greyson provides the best evidence that we may experience something after death. Dr. Greyson dedicates his book to "those who faced death and generously shared with me their most personal and profound experiences."

In the book, Dr. Greyson shares many of the remarkable stories he has collected over his 40 years of research into near-death experiences. Many of these people have suffered serious injuries during which their heart stopped, or they stopped breathing and were technically dead. When they were later revived, they recounted extraordinary experiences, such as feeling separated from their physical body, reviewing their life in great detail, or a feeling of unconditional love radiating from a being of light. Dr. Greyson created a scale of 16 of such features that he heard repeatedly in such reports to standardize a definition of near-death experience.

After reading that book, I have no doubt that near-death experiences are real. Greyson includes in his book references to Greek and Roman accounts of what appear to be near-death experiences. That historical record combined with the many recent and apparently authentic accounts of people who shared their experiences with Dr. Greyson suggest that many people, regardless of their religious views, have such experiences. What we don't know is whether these experiences are the product of a brain in the crisis of impending death or a connection with a spiritual realm that we don't understand.

As I see it, we are all seekers of the truth, whether scientists, Christians, Muslims, Buddhists, or Hindus, trying to understand the mystery of our place in the world. We should enjoy the journey together and respect and honor everyone's beliefs and traditions about death and the afterlife. Even if they are just sacred stories arising from the hopes and dreams of people struggling to confront the fear of death, they deserve our respect.

Despite my doubts about some of the central tenets of Christianity, I love much about the churches that I grew up in and was married in. They baptized me and my children and gave me comfort and fellowship for many years. The people whom I encountered in those churches were largely empathetic, sensitive, intelligent, and generous people trying to do good in the world. Those churches provide a place, music, ritual, and protocol for major life events, filling a void that few other institutions fill. When I imagine my funeral, it is in a church.

It is tempting to see the hand of God at work in my life. When I married Laura in 1983, neither of us imagined that she would become a leading scientist in the disease that threatens my life. The genetic mutation for my family's ALS was found in 2011, and pharmaceutical companies began to develop therapies to target that specific genetic mutation. My symptoms started in about 2012. Laura's connections resulted in my being enrolled in a clinical trial for a promising new drug. I ask myself, *Is God coordinating these events to give me a chance? Or is it simply random chance, or karma, or good luck?*

Asking questions is the beginning of knowledge. Answering these questions should involve application of our best, rigorous, rational thinking and the full breadth of scientific knowledge that we humans have assembled over time, as well as a healthy dose of tolerance and humility.

We must remember that we humans are limited in our ability to perceive the world around us. Even the animals we consider inferior to us far exceed us in some respects. Dogs have a far superior sense of smell and hearing. Bats and dolphins use echolocation, finding their place by bouncing sounds off of objects in their environment. Some birds are able to navigate thousands of miles in flight to specific migratory destinations without GPS. Octopuses have arms that are really an extension of their brain, highly sensitive appendages that seem to think on their own. There is a lot we humans don't know even with all our scientific instruments.

Why can't we say that we don't know, that our ability to perceive the world around us is not sufficient to provide evidence of such a God? Perhaps because our intuition tells us that there should be a God, and perhaps because the Christians, the Muslims, the Hindus, and others say there is a God. Perhaps it is fear of what others will think of us.

Given all my life experience, it seems to me that the idea of God, as an omniscient creator presiding over the universe, is an attempt to fill the gaps of what we don't know. Believers assert that there is a God because it's an easy answer to many of the questions we can't answer scientifically.

Instead, I believe that our perception of this world is not complete. I don't know if my conscious awareness will expire when my body dies, or if there is something beyond this life. I am humble enough to say that I don't fully understand the truth of our existence on this earth and skeptical enough to doubt all of those claiming to have the answers. I embrace the mystery and leave it to you to draw your own conclusions.

Chapter 12:

Pharma

Waves of change in the world impact our lives in ways big and small. Wars, the industrial revolution, the digital revolution, even the nationalist populism that brought us Brexit and Trump, are far-reaching societal changes over which we have little control but which significantly affect us and our families. The sweeping change that has occurred over my life, and which may save me, is what I call the genetic revolution.

Many trace the beginning of modern molecular biology to 1953 when American biologist James Watson and English physicist Francis Crick first described the double helix structure of DNA. Like most scientific discoveries, their work was based upon the work of many others, including most notably Rosalind Franklin, whose crystallographic evidence of the structure of DNA was shown to Watson and Crick by her estranged colleague without her consent. Watson and Crick won a Nobel Prize for the achievement in 1962, and the lack of any credit for Franklin in the story is viewed by some as an egregious example of the sexism of the time. The role of women in science has improved, but it is still a field dominated by men, and those women who succeed, like Laura, are extraordinary.

The discovery by Watson and Crick (and Franklin) led to rapid insights into the genetic code and how genes are translated into proteins. During the 1970s and 1980s, scientists began to map genes to the chromosomes and discover how variations in genes were related to certain diseases or how patients responded to drugs. When my mother was diagnosed with ALS in 1989, much was still unknown about ALS. Although we knew the disease had a genetic origin because of the number of affected individuals in our family, we didn't know which gene was involved. At that time, there were no drugs approved to treat ALS. Doctors working with my mom could only attempt to address her symptoms. They gave her specially shaped utensils designed to be easier to grip.

The lack of drugs for rare diseases was the problem Congress was seeking to address when it passed the Orphan Drug Act in 1983. Pharmaceutical companies weren't likely to invest in research and development funds if the disease was rare and the patient population small because that combination of factors made it hard to make a profit. The Orphan Drug Act, which provides incentives for drug companies to develop drugs for rare diseases, has been controversial but is largely viewed as a success. The act defines a rare disease as one that affects fewer than 200,000 people in the United States. About 20,000 Americans are living with ALS at any given time, so ALS qualifies as a rare disease under the act. The act gives drug manufacturers seven years of market exclusivity for a qualifying drug, tax credits equal to 25% (reduced from 50% in 2018) of the development costs for the drug, waiver of certain FDA user fees, and assistance in the drug approval process including fast track approval.

The act has been effective in incentivizing development of drugs for rare diseases. In the decade before the adoption of the act, only 10 drugs were approved by the FDA for the treatment of orphan diseases. Since the adoption of the act, over 600 drugs have been approved for treatment of orphan diseases. The success of the Orphan Drug Act in the United States led to similar legislation being passed in other countries, most notably Japan in 1993 and the European Union in 2000.

Critics say that the pharmaceutical industry is taking advantage of the act to charge outrageous prices for drugs, and politicians are increasing their calls for some controls on drug pricing. Industry analyst EvaluatePharma's Orphan Drug Report 2018 reports that the average cost for an orphan drug in 2017 was $147,308 per year. In an opinion piece published on May 14, 2018, *New York Times* columnist Paul Krugman wrote that the reason we haven't allowed Medicare to negotiate drug prices with Pharma is that "Pharma has bought itself enough politicians to block policies that might reduce its profits."

One example hits close to home for me. Recall that the drug that I have circulating in my spinal column and brain as I write this is an antisense oligo (ASO), developed by Ionis Pharmaceuticals in collaboration with Biogen. Another ASO developed by Ionis in collaboration with Biogen is Spinraza, which was approved by the FDA on December 23, 2016, to treat spinal muscular atrophy (SMA). Spinraza is a miracle drug for families with children affected by SMA. Children born with SMA previously suffered with progressive degeneration and had a drastically shortened life expectancy, most dying by age two by suffocating. With Spinraza, however, babies born with the disease who are treated pre-symptomatically develop into normal, healthy babies.

It would be a better story if we didn't have to talk about the cost. Biogen charges about $750,000 for the first year of treatment with Spinraza (six treatments at $125,000 per treatment) and $350,000 for subsequent years. Despite the high cost, I suspect insurance companies will have a hard time denying coverage for babies with SMA because the disease affects only one in 8,000 to 12,000 infants and, for heaven's sake, they're babies! To deny coverage would be a public relations disaster.

If the drug that Biogen is testing on me works as well as Spinraza, I imagine it may be priced in a similar range. Will the insurance companies and Medicare cover that kind of cost for less sympathetic old codgers like me? If not, it won't take long for those prices per year to eat up our retirement savings. Some drug companies provide certain drugs free or at a reduced cost for patients whose insurance will not cover the cost, and maybe Biogen will do that. In any event, there are a lot of uncertainties ahead.

This issue of drug pricing is complicated. Drug development is a risky business, and there needs to be a profit incentive to motivate companies to take the risk of embarking on the challenging and costly journey of developing a drug. A 2018 study by the Tufts Center for Drug Development estimated that it costs drugmakers $2.6 billion to develop a new prescription medicine that gains marketing approval. That estimate includes costs for the seven new drugs that fail along the way for every new drug that gets approval.

So the sales projections for any new drug need to be sufficient to absorb that cost and provide a profit. If the number of anticipated patients is small, the price has to go up. Still, the pharma companies appear to be doing well, based on the number of pitches I see on TV.

The first drug to treat ALS, Riluzole, was approved by the FDA in 1995. I take a Riluzole pill every day. Riluzole does not slow muscle deterioration or improve symptoms, but studies showed that it could slightly prolong the lives of patients. How much additional time does it give you? About three months. I assume that's some kind of average, though, and it may be that the effect is greater for patients like me whose symptoms have not advanced too far. But for us ALS patients, even three months is important. I'm looking for anything that helps.

Riluzole works by decreasing glutamate levels in the brain and central nervous system. Glutamate is the most abundant of several amino acids that act as neurotransmitters between nerves or between a nerve and a muscle, and in normal individuals, glutamate is released to transmit a signal and then is quickly cleared away. In ALS patients, glutamate is not cleared properly, and excess amounts accumulate in the synapses around nerves, causing prolonged excitation of nerve cells, which damages the cells. I imagine this is responsible for the involuntary muscle twitching, or fasiculations, that I occasionally experience. Clearance of the glutamate is the job of the glutamate transporters, one of which is excitatory amino acid transporter 2 (EAAT2). Dysfunction or reduced expression of EAAT2 has been found in many neurodegenerative diseases.

You might be thinking "glutamate, glutamate…isn't that a food additive?" You'd be right. Monosodium glutamate, or MSG, the stuff that some people worry about in Chinese food, is a form of glutamate. Does eating MSG increase the level of glutamate in the nerve synapses of your spinal cord and brain? Some people say MSG makes them feel jittery. Too much glutamate?

The second drug to treat ALS, Edavarone, was approved by the FDA in 2017, 22 years after Riluzole was approved. I am not taking Edavarone, in part because I would be excluded from the Ionis/Biogen ASO study if I were. Biogen permits study participants to take Riluzole, however. Also, the current treatment protocol for Edavarone is demanding. It is administered in 28-day cycles. During the first 14 days, physicians inject Edavarone into the patient's vein via an IV infusion for about 60 minutes every day. During the second 14 days, the patient is drug-free. During the third 14 days, physicians inject Edavarone in the same manner for 10 of the 14 days, followed by 14 days drug-free. Then the cycle is repeated.

Edavarone is a free radical scavenger. You've heard about free radicals (also called "oxidants") from TV commercials for various foods that claim to be antioxidants, like free-radical-fighting pomegranate juice. Free radicals are problematic because they have an extra electron that causes the molecule to be unstable, likely to wreak havoc by binding where it causes damage. Free radicals are also a normal byproduct of the process by which our bodies convert food to energy, so we are always producing them. In healthy individuals, there is a balance between the amount of free radicals produced and the amount of antioxidants floating around to soak them up. We get antioxidants by eating fruits, vegetables, and lots of leafy greens, so a good diet is important. Even so, some believe that normal aging is the accumulated wear and tear of free radicals doing their damage. Of course, normal aging in Dan

Buettner's Blue Zones, where people eat a lot of plant-based food, is a lot longer than in the United States.

Edavarone is based on the assumption that ALS patients have a lot of oxidative stress; i.e., lots of free radicals killing nerve cells, and that reducing the amount of free radicals will have a positive effect on disease progression. This theory is supported by the finding that mutations in a gene encoding an enzyme (Cu/Zn-binding superoxide dismutase, or SOD1) that removes free radicals have been found to cause about 10-20% of all cases of familial ALS and about 1-2% of cases of sporadic ALS. Curiously, however, when scientists deleted the SOD1 gene in mouse models, the animals did not get ALS, suggesting that it is not the loss of the function of SOD1 that causes the disease. Nevertheless, in small studies of ALS patients who were recently diagnosed and whose disease had not progressed far, Edavarone was shown to be effective in slowing the disease progression.

The price for a year of treatment with Edavarone is reported to be $148,000 (about average for an orphan drug). Interestingly, it costs only $35,000 in Japan.

The fact that only two drugs have been approved for ALS in my lifetime tells me that ALS is a tough nut to crack. But the advancing field of genetics has given us a better foundation for finding effective therapies than ever before. Now we know that multiple genetic mutations cause ALS, so there may be different disease mechanisms that result in similar symptoms. That would suggest that treatments should be tailored to fit the particular genetic mutation. In other words, the treatment for an ALS patient with a SOD1 mutation is likely to be different than the treatment for an ALS patient with a *C9orf72* mutation. Currently, however, we all take Riluzole or Edavarone or both.

The future will be different. The Ionis/Biogen drug being tested in the trial that I'm in is designed specifically for the *C9* mutation that I have. Laura is collaborating with Biogen to evaluate another drug based on antibodies targeting the proteins created by the *C9* mutation and that seems to help my little mouse cousins. The team at Johns Hopkins is testing another ASO for the SOD1 mutation. I am happy to have a chance to be the beneficiary of the scientific progress that is taking place. I just have to hang in there until these drugs come to market. So far so good.

Chapter 13:

Laura

I met Laura at St. Olaf College when I was a senior and she was a sophomore. We were both on the cross country and track teams. Although the men's and women's teams had different programs with different coaches, both teams were at Skoglund Athletic Center at about 3:00 p.m. every day, and during the indoor track season, both ran workouts and track meets in the same fieldhouse. So there was plenty of opportunity for the men and women to see each other, if not talk to each other, and see each other in circumstances that left little room for pretensions.

Even before I spoke to her, I knew a fair amount about Laura. I knew that she had a beguiling smile, eyes that were at once wary and intelligent, and that she looked great in shorts. I knew that she was willing to run to the point of exhaustion and then smile and laugh with teammates who liked her.

So when she and some friends climbed through the fire-escape window into our basement apartment on St. Olaf Avenue where we were having a track party, and because I had had enough beer to get my courage up, I started talking to her. I probably shouted over the music (e.g., "Brick House," by the Commodores), "Why did you come through the window? We have a door." That was the start.

She was a biology major and I was an English major. I generally studied in Rolvaag Memorial Library, and by asking a few of my science buddies, I learned that Laura often studied in the Science Center. So one weekend, probably the weekend after the party, I went to the Science Center to study. Sure enough, Laura was there, affording an opportunity for a whispered exchange. I said, "Hey, you study here, too?" She later told me that she was on to my scheme.

Eventually, I asked her out. She had something going on that night. I tried again. Another conflict. I asked again. Another conflict. She must have given me enough reassurance to keep asking because the fourth time we made a date. We went out for pizza and to a movie in Northfield. Nobody had cars at St. Olaf, so we walked the mile or so from campus to Northfield's small downtown. The movie was *All the President's Men* with Dustin Hoffman and Robert Redford. Ironically, the U.S. political world seems to have returned to a similar state of dysfunction.

Initiating a romantic relationship with someone is a tricky business. We carry a lot of baggage into that dance: our parents and siblings; ideas about romance and sex from books, TV, and movies; and, of course, the relentless demands of biology. I was pretty sure I was looking for someone like Olivia Newton-John; Laura was probably looking for John Travolta. But Laura was fun to be around. She was and is beautiful, energetic, and optimistic. She is a left-handed science nerd who occasionally comes up with entertaining malaprops, like "I'm going to put my thumb down," conflating "under my thumb" and "put my foot down." In those days, both she and I were unsure of ourselves and trying hard to make our way in the world. We had a shared ambition to make something of ourselves and took comfort in each other's lack of arrogance and sense of humor.

But I was a senior, and she was a sophomore, so when I graduated, I thought the relationship might suffer. Shortly after graduation, however, a St. Olaf classmate, Pete Scheuer (sadly, Pete passed away from a heart attack in 2017), got a group of us together for a trip in the Boundary Waters Canoe Area in Northeastern Minnesota, and Laura and I went together. The BWCA is a magical place, if the weather is right and the mosquitos "aren't too bad" (that's how we Minnesotans say it). It is a wilderness area filled with glacial lakes, streams, and forests. A trip to the boundary waters involves canoeing across pristine wilderness lakes, portaging the canoes to the next lake, and canoeing some more until arriving at your campsite, then getting up the next day and doing the same. In June, which is when we went, the mosquitos are terrible. What I remember about the trip was the mosquitos, that Pete was too ambitious in estimating how much distance we would cover each day, and wishing that Laura and I had our own tent instead of the tent that we shared with a buddy of mine.

Looking back on that now, I ask myself, "Why didn't I just buy a tent?" Then I recall the obvious answer. I had no money. At the end of the school year, before whatever summer job I had lined up, money was very short, and an extravagance like a new tent was out of reach. Not that I'm complaining. I know kids now face much higher tuition than we did. I recall that St. Olaf, which I thought of as a fancy, expensive, private college, was about $4,000 in tuition and room and board by my senior year. I was able to cover that with grants, student loans, and the earnings from the summer construction jobs that my dad lined up for me. I knew money was tight at home, and so I didn't expect my parents to help. I'm sure they would have if I had asked, but they had no idea what their role was supposed to be. Neither of them had been to college, and my brother and I were the first of their four kids to go off to college. We pretty much handled the financial stuff ourselves, filling out the financial aid forms, deciding what we had to pay, and whether we had enough to swing it. I didn't feel poor, but let's just say it was a real treat to get a new pair of jeans. So a new tent for one weekend was not in the cards.

That summer I was living in an apartment near the University of Minnesota, where I planned to start law school that fall. Laura's family lived in Minneapolis, so we saw each other frequently. I was driving a motorcycle at the time, and Laura's mother hated to see us ride off with Laura on the back of that thing. I think Laura hated it too, but she wasn't ready to make the motorcycle an issue in the relationship. That would come later.

Laura's mom, Arvella, died the day before her 98th birthday on March 13, 2020. She had lived for the last years of her life in a memory care unit at a fine elder care institution called Boutwell's Landing in Stillwater, Minnesota. The daughter of a doctor who practiced in Great Bend, Kansas, she went to law school at a time when few women did so, but never practiced law. Her first marriage ended tragically with the unexpected death of her husband, and she married Laura's father, Page, at age 31. They were a handsome couple. Their wedding photo looks like an ad for a movie. For their 50th anniversary party, the kids blew up the photo to poster size and had the attendees sign it. That poster-size photo of their wedding day, with all the signatures from the anniversary party, hung in Arvella's room at Boutwell's Landing until the end, a reminder that we should be grateful for what we have today, because bit by bit, time takes it away.

When I started law school in the fall, Laura went back to St. Olaf for the fall semester of her junior year. In those pre-cell phone days, the 45 miles from Minneapolis to Northfield, where St. Olaf is located, was a bigger barrier than it is now, but we kept in touch by (landline) phone, letters, and occasional weekend

visits. In the dorms at St. Olaf, there was generally one
phone for every "corridor," which included maybe six
rooms. Typically, it was on the wall or at a desk in a
common area, and students took turns to use it. In my
case, at law school I was near a phone only when I was
at home in the apartment, which wasn't often since I did
all my studying at the law library. So the phone calls
were not frequent. And there was no voice mail. The
best you could do was get someone to jot down a note
as to who had called and when. Laura was planning to
go to Costa Rica for the second semester on an
independent study in biology, and I was plowing my
way through the first year of law school, so although we
were both excited about the blossoming of new love, we
were busy, and neither of us could have imagined the
future that lay ahead.

The semester in Costa Rica demonstrates Laura's
boldness and the quality of the science program at St.
Olaf. She could have studied abroad like most students,
going somewhere with a structured program, maybe at
a foreign university, but instead she chose to do an
independent study through a program run by the
Associated Colleges of the Midwest. Another woman
on the track team had traveled to Costa Rica with the
ACM program two years before and gave it a good
report. So Laura worked with her advisors to develop a
study to examine the cholesterol levels of middle-aged
men in Costa Rica. She traveled with a group of students
who upon arriving in Costa Rica had a month of
orientation before splitting up to go to their respective

sites. Most of the students went to the biological research station in a cloud forest called Monte Verde. Laura wanted to pick a site where she would live with Costa Ricans and learn to speak Spanish, so she decided to work under the supervision of Dr. Leonardo Mata in Puriscal, Costa Rica.

Dr. Mata was an accomplished physician who had done graduate work at the School of Public Health at Harvard. He was good-looking and intelligent, and, based on the stories that Laura told me, had complicated views about women—arising, I suppose, from the machismo culture in Costa Rica, his studies on breastfeeding, and the pretty, young female students looking to him for guidance. He never touched Laura inappropriately, but at times his conversation would drift into explicitly sexual topics, unnecessarily, under the cover of health topics. Or, he would use innuendo: He once asked Laura if she wanted a banana, then followed up with "Do you prefer the big American bananas, or the smaller, but sweeter, Costa Rican bananas?"

After Laura had returned from her trip in 1982, I suspected the stories about Dr. Mata were exaggerated a bit. But a trip with Laura to Costa Rica in 1987 convinced me that I was wrong. First, when Laura called Dr. Mata to say we were in the country and would like to meet, he asked her if she could come alone. When she demurred, he invited us both to a presentation that he was giving and suggested that we chat afterward.

During the presentation as he was going through his slides, suddenly there appeared a photo of a man and a woman having sex. It was a complete non sequitur. He said something like "How did that get in here?," smiled, and continued. During the conversation afterward in his office, not five minutes had passed before he was talking about the sexual practices of gay men in a conversation that had started with his work on HIV. I left convinced that Laura had not exaggerated.

Despite Dr. Mata's eccentricities, Laura had a great experience in Costa Rica. She learned how to draw blood from people, a skill she has continued to use throughout her career to obtain blood samples for testing, including blood samples from my mother and me to bring to the lab. She learned to speak Spanish. She has kept in touch with the families she stayed with, and we have visited them in Costa Rica several times over the years. She has even organized two scientific meetings in Costa Rica.

There were a lot of reasons not to go on a trip like that. She started with only a rudimentary knowledge of Spanish. She would be traveling and working without friends from St. Olaf. She had to design and execute her own research study. She had to find a family to take her in as a boarder. But even as a 20-year-old kid, Laura was willing to throw herself into new challenges, often without knowing if she would sink or swim, but with optimism that it would be an adventure worth having.

I loved that adventurous spirit, which, surprisingly, is combined with a rigorous approach to personal safety that I fully understood only as our relationship matured and she was less guarded about expressing her opinions. I confess that when I first started dating Laura, I didn't always put my seatbelt on before starting to drive, but Laura soon trained me that the car doesn't move until the seatbelt is on. I grew up in a rural area where we never locked our house. For Laura, house doors are always locked. The motorcycle was the subject of an increasingly frequent commentary about the dangers of motorcycles. And storms. Laura is both fascinated and horrified by severe weather, especially tornadoes.

The root of her emotions about tornadoes lies in a childhood experience. When she was five years old, a tornado was spotted near her home in South Minneapolis over Lake Nokomis. Arvella rounded up the kids and ordered everyone to the basement. Laura's father, Page, refused and instead opened the back door to take a look at the sky. Arvella made a scene trying to get Page into the basement, making quite an impression on young Laura. So Laura has an aggressively proactive approach and high level of concern about tornadoes, or, now that we live in Florida, hurricanes, or any severe weather. She is an enthusiastic user of the weather apps on her smartphone. She can be mesmerized, however, watching the Weather Channel footage of storms in other places.

So even when we were young, Laura was busy saving me from car and motorcycle accidents, robberies and storms of all kinds, especially tornadoes. Now the stakes are higher, and she is trying to save me from ALS.

When Laura returned from Costa Rica in the spring of 1981, we took up where we had left off. Hanging out, movies, parks, bike rides—we had become best friends, and that summer, lovers. We had been dating for about 18 months. I was 23, and she was 21, but we both felt secretive and a little guilty about having sex. It was only a few years after *The Mary Tyler Moore Show* had begun to let viewers know (gasp!) that Mary was having sex as a single woman at 30. There was still a lingering expectation among families like ours that you didn't have sex until you were married. We never spent the night together or enjoyed the simple joy of waking up in the same bed. And living together was simply out of the question. So I'm sure there were some raised eyebrows when we announced that we were going on a week-long hiking trip together in Glacier National Park at the end of the summer. The only thing that made this arguably acceptable was that my law school buddy, Jim Ladner, would be joining us.

Jim is a great guy and was more experienced at hiking than either Laura or I. He advised us to leave behind anything that we didn't need and bring only the essentials. So again, there was only one tent for the three of us. But by the end of that first 14-mile day, we understood that every ounce in the backpack counts heavily toward your weariness at the end of the day. Halfway through that first day, Laura discovered that her new boots chafed her feet, and so she swapped them for her running shoes. I carried the boots tied to my backpack the rest of the trip. She has been trying to make it up to me ever since.

We felt our legs and backs grow stronger every day. All around were dazzling, snow-capped mountains, glacial rivulets, and sparkling lakes in the valleys. Along the way we saw a juvenile grizzly bear on the trail 25 yards ahead of us. We stopped, frozen, until Jim started to get his camera out (I'm not sure that was the right move). The bear got up on his hind legs, looked at us, then got back on all fours and went off into the forest.

We hiked out on day six feeling like we had done something grand and celebrated with a pizza and beer.

After Laura's graduation from St. Olaf and during the summer after my second year of law school, we had been together over two years and we loved each other, but...I was starting to think the train was on the fast track to marriage, and I started to get cold feet. I'm not proud of it, and I realize I almost derailed the best thing that's ever happened to me, but I told Laura that I wanted to take a break. I wanted to date other people. Looking back on it now, I say to myself, *You egotistical bastard, how could you do that to her?*

I was and am not a risktaker. I overthink things. It makes me a good lawyer but a lousy lover. So I told myself I wasn't sure and that I should go on some more dates to figure some things out. I made a date with another law student who had caught my eye. The first date was so unsuccessful that I thought surely a second would be better. It was not. She was pretty, and nice enough, but didn't have Laura's energy or sense of humor or sparkle. Also, the awkwardness of dating someone new made me realize that this process was going to be terrible.

I worried when I heard that Laura had gone on some dates with someone else.

It was less than a month before we were seeing each other again. What I didn't know then is that there are really very few people with whom I would be happy as a spouse. That indefinable chemistry that draws one person to another occurs rarely. That chemistry, that spark, is essential but not sufficient. The spark has to be nurtured and protected, and when it grows to a fire, the fire has to be tended to continue over time. In other words, love is a choice. We have evolved to seek out and find a mate, but good relationships are built over years of kindness and mutual support. With Laura, I knew there was a spark (hell, a pretty good fire), but I walked away from the fire. Thankfully, I wasn't gone long enough for the fire to go out.

There was never any big proposal on bended knee. We were open with each other, and so we talked about the future from time to time and tentatively approached the topic of marriage. We were burdened by the conventional expectation that the man should make the proposal, and so thinking that a proposal should be reserved for a special "moment," I carefully avoided saying anything that sounded like a proposal, and Laura avoided it too. But I remember talking with Laura for hours during a long drive sometime during the summer after I had graduated from law school and taken the bar exam. I had a job lined up with a law firm for the fall. During that conversation we decided to get married. And we decided to do it quickly. We wanted to live together.

If I allowed myself regrets, which I do not, I would say, "Why didn't you surprise her with a ring in some grand gesture like Paul did with Susan?"

Sadly, that wasn't me then. Paul is a better man. Paul was in graduate school and Susan was in medical school when they became engaged. Despite their busy schedules, Paul arranged for a photographer to follow them secretly at a hot air balloon festival while he tried to casually direct Susan to a good spot for the proposal. When the time and place was right, he dropped down on one knee and proposed. Susan was completely surprised. The photos of the visibly emotional couple with gigantic, colorful, hot air balloons filling up in the background were magnificent.

My consolation is that however lacking our proposal may have been, the marriage itself has been magnificent. Also, every parent wants their kids to do better than they did.

We made the decision to get married in late September of 1983 and decided that the wedding would be between Christmas and New Year's Eve. We liked the idea of a Christmas wedding and assumed that many of our friends would be in the area with their families for the holidays. The challenge of finding venues and planning with only a few months' lead time did not occur to us. Fortunately, we had Arvella on our side. We told our parents shortly after we decided, which happened to be only four days before Page and Arvella left for a trip to China. Traveling to China in 1983 was a more exotic thing than it is now, and they had been planning for that trip longer than the time we had allotted to plan for the wedding. But in those four days, Arvella and Laura booked the church and a reception location, got a dress for Laura, and bought fabric for the bridesmaids' dresses (each of the bridesmaids was free to choose her own design).

At the time, I was blissfully unaware of all the planning for the wedding. My obligation was simply to show up on the appointed day and get hitched. Now, having seen the efforts involved in several weddings for nieces, nephews, and my son, I know what goes into an event like this, and I recognize what a naive rube I was.

As the date, December 29, 1983, approached, a cold front moved into the twin cities. The average temperature in Minneapolis for the month of December in 1983 was 3.7 degrees Fahrenheit, a record low. On December 19, the temperature dropped to -29 degrees Fahrenheit. I was concerned that they wouldn't be able to heat the church, St. James Episcopal in Minneapolis, which is old and inefficient with the typical high nave that I imagined would have to be filled with warm air before any got down to where the pews were. I was relieved that the temperature had warmed up by the time of wedding to -4 degrees.

Despite the cold, the church was warm and festive with decorations for Christmas. The guests arrived to carols sung by the church choir. We had two officiants: Bruce Whitmore, Laura's brother (who was at that time an Episcopal priest), and Morris Sorenson (a Lutheran pastor and the father of my sister-in-law, Mary Ranum). We should have hired a better photographer than Bill Miller, a friend of the Whitmore family, whose confidence as a photographer exceeded his skills. He did, however, work for free. The reception was held at the Gale Mansion, owned by the American Association of University Women, of which Arvella was a member, a lovely 1912 mansion on the National Register of Historic Places.

So thanks to our family and friends, our wedding came together quite nicely. The honeymoon, not so much.

Since Laura was busy with the wedding preparations, we agreed that I would make the reservations for a modest honeymoon trip to the north shore of Lake Superior, a popular vacation destination for Minnesotans. We planned to relax, cross country ski in the parks along the way, and enjoy sitting by the fireside toasting our new life with a glass of wine. When I began to call around, however, I discovered that the week between Christmas and New Year's is heavily booked at the hotels on the North Shore. I imagine that I was using some book or magazine for information because the internet with all of its pictures, reviews, and information was not yet available. So with imperfect information, I found a place that had a room for the first two nights, December 30 and 31 (we had stayed in Minneapolis on the 29th), that looked nice enough.

When we arrived, however, we discovered why people were not clamoring to stay there. Not that there was anything really wrong with the place; it was just undistinguished. It was a typical roadside motel—not the place for the first two nights of your honeymoon. We arrived late that first night. There was no dining room, and it was too late to drive to a restaurant, but the motel had a little snack area where guests could buy frozen pizza from a machine and heat it up in a toaster oven. Laura is a good sport, and I remember eating that pizza and laughing at ourselves, secure in the knowledge that as long as we were together, where we were didn't matter so much.

After New Year's Eve, we had reservations at a more upscale place that we appreciated all the more because of the place we had just left. Now, 38 years later, the honeymoon has become one of those stories in a marriage that becomes symbolic, a cautionary tale of the dangers of insufficient advance planning for travel. As a result, we became better travelers, and whenever we book a room at an extravagant hotel, which we have done all over the world, I justify it as a reparation to Laura for my honeymoon failure.

In the years since then, Laura has become a serious scientist and is making significant contributions to her field, including particularly the discovery of RAN Translation, which may be the primary driver of my ALS. This serendipitous connection between Laura's work and my disease deserves closer scrutiny.

In 2006, while studying the repeat expansion mutation in spinocerebellar ataxia type 8 (SCA8), Laura and her lab discovered that the mutation produces two RNAs instead of just one. This double-the-trouble scenario was also found in myotonic dystrophy type 1 and is now known to occur for many repeat expansion disorders.

In January 2011, Laura published a paper showing that these repeat-expansion mutations also break the previously established dogma that a start signal embedded in the genetic code is required to make proteins. In genetic code, that start signal is AUG.

Surprisingly, she showed that when a repeat expansion mutation is present, proteins are produced without the AUG start signal and extra, unexpected proteins are made – up to six for each expansion mutation. She called this aberrant translation process repeat associated non-AUG (RAN) translation and the surprising proteins produced RAN proteins. This discovery was so different and surprising that many scientists dismissed it.

The bombshell was dropped later in 2011, when scientists at the Mayo Clinic in Jacksonville and the National Institutes of Health published that an expansion mutation in *C9orf72* is the most common genetic cause of familial ALS, and as we subsequently learned, also the cause of my ALS. Laura suspected immediately, and confirmed within months, that the expansion mutation in the C9orf72 gene produced RAN proteins.

All of this meant that Laura's decades of work on a different disease suddenly became relevant to the disease in my family, and as we would subsequently learn, to me. She threw herself into learning everything she could about this mutation. She and her lab created a mouse model of the disease by modifying the DNA of these mice to have the *C9orf72* mutation from my family. She began to explore potential therapeutic options, including metformin.

Whether due to serendipity, or the hand of God, or the unknowable synchronicity of the universe, I am awed by and grateful for my wife's knowledge of and expertise in my disease. When Lou Gehrig, the famous New York Yankees baseball star, left baseball because of his advancing ALS, he said in his farewell speech, "I consider myself the luckiest man on the face of the earth." I feel the same way because the love of my life also just happens to be an expert on the disease that threatens me.

Chapter 14:

Wellness

I once heard that we should all try to die young as late as possible. I like that idea. No one wants to grow old. We just want as much quality time on this earth as we can get. While we all like the idea that science will come up with a pill to fix what ails us, by our 40s or 50s, most of us have a pretty good idea of how to stay young. Eat right, exercise, get enough sleep, avoid stress, and surround yourself with people you love. The art of living is doing all of that well.

When Laura and I realized that my voice was deteriorating and suspected ALS, we knew that there were really no effective treatments for ALS. So we decided to do whatever we could do *now* to slow or even stop the progression of the disease—in other words, to keep me as young as when I was free of the disease. During a trip for a scientific meeting, Laura was flipping channels in her hotel room when she happened upon a public TV program featuring Dr. Joel Fuhrman and his "Eat to Live" diet. Soon we had the book, *Eat to Live*, and were converts to the diet, which is a challenging vegan diet, but it does allow legumes, like beans, peas, and lentils.

Dr. Fuhrman's argument is unassailable: The typical American diet, high in fat and refined carbohydrates, is causing obesity and the myriad health conditions that follow, including diabetes, hypertension, heart disease, etc. He discourages the consumption of animal products like meat and dairy because of suspected links to cancer and high cholesterol. The most striking evidence for this view is the China-Cornell-Oxford Project (also known as the China Project) led by Dr. T. Colin Campbell of Cornell, which studied isolated populations in China who ate a primarily plant-based diet and compared them to nearby populations with a diet including meat and dairy. At the time the study was conducted, the populations were very stable. For the most part, people did not travel and lived their entire lives in the same village. Fuhrman writes, "Researchers found that as the amount of animal foods increased in the diet, even in relatively small increments, so did the emergence of cancers that are common in the West. Most cancers occurred in direct proportion to the quantity of animal products consumed."

Even though cancer was not my problem, we embraced the diet on the theory that we needed to be as healthy as we possibly could to combat my disease. We found an Indian restaurant that had a great vegetarian curry that we would enjoy over rice with garlic naan, vegetable samosas, and a Taj Mahal beer. We went to an Indian store and bought spices and started to cook more Indian recipes. The spices helped to distract us from the savory umami flavor that we missed from the meat.

When social circumstances offered an excuse to cheat on the diet, I would grab the chance to enjoy a burger or a steak, until another health issue arose. One evening when Laura was out of town, a friend invited me to dinner to celebrate his birthday, and the meal included grilled steaks and, to make things interesting, snails. I had never eaten snails, but I tried one, concluded one was enough, and then dug into my steak, baked potato, and asparagus. So when I woke up that night sweating with itchy hives all over my body and wondering, *What the hell is happening to me?,* I concluded that I was allergic to snails. As I lay there, scratching my head, I began to feel a tightness in my chest, and I thought I was having a heart attack. I decided I should get up and get an aspirin. So I walked into the kitchen, grabbed a glass to get some water, then fainted and woke moments later on the floor with a broken glass beside me. This was not good. *Just get to bed to lie down,* I told myself, so I carefully got up, made my way to the bedroom, and lay down. I eventually fell asleep. When I woke up, I was fine and vowed never to eat snails again.

Months passed. We ate a lot of black beans and rice, lentil soup, Indian curries, salads, and smoothies, and avoided processed foods, bread, pasta, pizza, dairy, and meat. At this time, I was traveling to Minnesota every month to check in with my law firm, and during one of my visits, friends took pity on me staying at 2116 Carter Avenue, St. Paul, all alone and invited me to dinner. Again, the menu was steaks on the grill, and I enjoyed another forbidden steak with salad and a glass of red wine. Again, I woke up in the middle of the night in distress, hives all over my body. As I lay there, I decided it had to be the steaks since this had happened months apart, each time after eating a steak. Since I had been through this before, I just lay there and eventually went back to sleep.

In the morning, I got out my computer and Googled "beef allergy" and I pulled up several articles that associated a red meat allergy with tick bites. Because of my encounters with ticks, I knew immediately that I was wrong about the snails and that I had a red meat allergy caused by tick bites.

Our home in Florida is about three miles from the San Felasco Hammock Preserve State Park, and Laura and I went there frequently in the years after we moved to Florida to run (back in the days when we were still running) or walk on the trails through the woods. It is a beautiful, natural place where the trees offer a welcome shade from the summer sun and the breezes are a little cooler than in the city. We've seen deer, armadillos, turtles, a variety of birds, and even a bobcat while walking in the park. I almost stepped on a coral snake there, which caught my eye because of its striking colors. I studied it carefully and then looked it up on the internet to identify it after we got back, where I learned the rhyme "red touches yellow, kills a fellow."

We also occasionally found ticks on us after our walks. Especially me, since, we think, my hairy legs give the ticks more opportunity to latch on as we pass by. On one of our visits to the park, a path that we had walked down in the past was closed, and we made what in retrospect was the bad decision to walk down it nevertheless. The problem was that the grasses and other plants had grown tall enough to reach our knees. We didn't give it much thought as we waded through the grasses, but when we got home, we found that swarms of larval stage ticks had attached themselves to my legs. These were tiny ticks, only slightly larger than the period at the end of this sentence. I don't recall whether Laura had worn pants or shorts, but regardless, she had a fraction of the number of ticks that I had. We painstakingly searched for the tiny ticks and pulled about fifty of them off, but inevitably we missed some, and I found additional ticks in the days following because each bite raised a little itchy welt on my skin.

So, needless to say, I remembered my encounters with the ticks when later I read that tick bites can cause a red meat allergy. I also read that this is more common in the Southeast, where the lone star tick is common. When the tick bites a deer, rabbit, or other mammal it ingests molecules of alpha galactose, a sugar common to mammals except humans and monkeys. If the same tick later bites a human, it can transmit those molecules of alpha galactose combined with saliva from the tick in a way that causes the immune system of some people to identify alpha galactose as a foreign antigen that needs to be attacked. Then when that person eats a steak, alpha galactose is released as the meat is digested hours after the meal, and the immune system mounts an attack on the invading alpha galactose.

I have read that some people with the alpha gal allergy eventually recover the ability to eat meat after several years provided they avoid subsequent tick bites, so we have stopped going to San Felasco as frequently, and when we do, we cover our legs and spray ourselves with insect repellent. On my last visit to the allergist, I learned that my alpha gal antibody level was down significantly, but not enough for me to start eating red meat.

I suspect that my alpha gal allergy is related to my ALS in some way. Maybe my immune system is revved up by fighting the RAN proteins that are produced by my *C9* mutation in a way that has made me more susceptible to the alpha gal allergy. But that's another biological mystery story.

Since we were no longer eating meat, the alpha gal allergy really wasn't much of a problem, until we got a couple of additional books and reconsidered our vegan approach. Laura had heard from some of her scientific colleagues about a doctor at the University of Iowa, Terry Wahls, who had essentially reversed her multiple sclerosis with a diet based on paleo principles. We were interested because like ALS, MS is a disease of the nerves and muscles. We also discovered *The Paleo Approach*, by Dr. Sarah Ballantyne, by browsing in the diet books section of Barnes & Noble. Both Drs. Wahls and Ballantyne recommend organic, grass-fed meat, wild-caught meat and seafood, as well as organ meats and offal, as good sources of many nutrients and proteins.

The other way that Drs. Wahls and Ballantyne differ from Dr. Fuhrman is that they discourage consumption of any grains, beans, and legumes and also nightshades (potatoes, tomatoes, eggplant, bell peppers) because of their relatively high carbohydrate load and lectins. Lectins are a class of proteins, the most notorious of which is gluten found in all wheat products, that damage the lining of the gut leading to what has become known as a "leaky gut." Dr. Ballantyne's book goes into great detail about how damage to the lining of the gut and the resulting permeability of the gut may be at the root of many diseases. When partially digested fragments of these lectin proteins leak through the lining of the gut they can cause an immune response that may be responsible for celiac disease, irritable bowel syndrome, ulcerative colitis, Crohn's disease, and other autoimmune disorders.

Based on the *Wahls Protocol* and *The Paleo Approach*, we traded in our whole grains and beans and lentils for farmer's market chicken and turkey and seafood from a great local seafood store called "Northwest Seafood." In my opinion, it was a step up in our quality of life. We love to pair our big salads with a roasted chicken, covered in garlic, or grilled salmon, seared with stripes from the hot grill. The roasted chicken provides for several meals, and after we've picked the bones clean of meat, we throw the remaining carcass, bones, leftover skin, neck, and drippings from the pan into a crockpot with water and simmer it for 24 hours. We strain the broth with a cheese cloth, toss the bones and carcass, and use the resulting stock for soups, or just straight in a cup as bone broth.

Based on *The Wahls Protocol*, we don't limit our fat intake, although we do try to eat the right kinds of fats, like extra virgin olive oil, coconut oil, and lard that we render at home from pork fat from a small farm that raises their pigs in a pasture under live oaks (we've been there). Curiously, eating pork fat doesn't trigger my alpha gal allergy. Maybe there is less or no alpha gal in the fat as compared to the meat. We like to make creamy smoothies with greens like spinach, kale, or romaine and fruit like bananas, peaches, blueberries, strawberries, and coconut milk. It's best if some of the fruit is frozen, and we usually throw in some ice as well.

Dr. Ballantyne's emphasis on gut health confirmed all that I have been hearing over the last seven years of reading and listening to health and wellness media about the importance of the gut and the microbiome, the community of microbes in our intestines that works synergistically with our biology to digest our food. People even talk of the "Gut-Brain Axis" connecting gut health to chronic brain conditions like dementia and neurodegenerative disorders like Parkinson's and ALS.

I am convinced that getting the gut and microbiome in order is a foundational aspect to good health.

To improve our gut health, Laura started making kefir, kombucha, and sauerkraut. We found some soda at a grocery store that came in glass bottles with swing-top resealable caps and bought it for the bottles (we poured the soda down the drain). Now those bottles are typically lined up on our kitchen counter filled with fermenting kefir and kombucha. The color is various shades of red from the pomegranate juice we add to the mix. We like to pour a glass with half kefir and half kombucha for a delightful fizzy, tangy drink that is a little different each time. We also make sauerkraut by chopping up a head of red cabbage, then massaging it with salt until it gets soft and releases some moisture. Then we pack it into wide mouth jars, put in a glass "stone" to hold it down beneath the level of fluid, and seal it with a special cap that allows the gas to escape. After two weeks of fermenting we have a wonderful sauerkraut that is a tasty and colorful addition to our salads.

Although I enjoy the food we're eating on the Paleo diet and I believe it does have health benefits, I do miss many of the forbidden foods. My mom was a good cook and baker, and growing up we ate traditional Midwestern meals with meat, potatoes, and a vegetable, but there were also a lot of pies, cakes, and cookies. Like a traveler in a foreign land thinking fondly of home, I sometimes long for a slice of apple pie fresh from the oven, the smell of cinnamon wafting through the air, with a scoop of vanilla ice cream melting at the point where it rests against the flaky crust of the pie, or the cinnamon rolls with vanilla icing, or the chocolate chip cookies with walnuts. I confess, I have a sweet tooth, or had a sweet tooth. So on vacation, or on special occasions, I will indulge in a decadent dessert and enjoy it all the more for the rarity of the event.

The wellness community seems to be growing. I have enjoyed listening to the podcast "The Broken Brain" in which the silken-voiced host, Dhru Purohit, interviews various leaders in a variety of fields relevant to wellness and brain health (the podcast is now called "The Dhru Purohit podcast"). Dr. Mark Hyman produces the podcast and has also promoted a couple of Broken Brain "docuseries" that I've enjoyed listening to. I've learned a lot, but the take-home message is this: Eat right, exercise, get enough sleep, avoid stress, and surround yourself with people you love. Even if it doesn't cure you, it will make you happier.

A great barometer of how well you're doing on that simple directive is to listen to your body. I also pay attention to how Laura is doing, since she's eating pretty much the same thing as me and she's a little more sensitive to various foods. I know, of course, that we're different people and we may have different sensitivities to different things, but it may also be that what upsets her gut may also be bad for me even though I don't have an equally adverse reaction. We're both doing great on our Wahls/Ballantyne Paleo diet. Before COVID arrived, Laura traveled a lot, and when she's away from home it was harder to stay on the diet, and she often felt the effects of eating carelessly and came home with new motivation to stay on the diet. My gut is pretty happy most of the time regardless of what I eat, and I do enjoy cheating on the diet on vacation or on special occasions. I'm sure that our efforts to "eat right" will continue to evolve as we learn more, test recipes, evaluate the progress of my disease, and consider our quality of life.

As for exercise, we've been swimming a lot lately. When we moved to Florida in 2010, we bought a house that was originally built in 1994. Although the house was designed for a pool, with lots of windows and big porches in the front and back, the original owners did not build a pool. After my diagnosis in 2016, we began to think about building a pool and reevaluated our retirement savings in light of the fact that I might not live as long as we previously anticipated.

By July 2017 there was a big backhoe parked in the backyard digging a hole for the pool and dump trucks trundling back and forth through our previously quiet neighborhood hauling dirt and fill. Laura and I had done a fair amount of swimming in the past (we did triathalons for a few years in the 2000s), so we built a lap pool, 25 yards long and two lanes wide. Our designer, Elyse Ostlund, created a design that integrated a spa, an outdoor kitchen, and the porch into a beautiful outdoor space beside the rectangular pool. Elyse's husband, Mark, built a pergola over the outdoor kitchen. Our pool contractor, Mike Flanagan of Pools & More, was great to work with, always reasonable, and willing to solve every problem. When we started the project, we heard some horror stories about bad pool contractors, but Mike was responsive, professional, and a pleasure to work with. Maybe it's because of the construction work I did with my dad as a kid, but I greatly admire people like Elyse, Mark, and Mike who can take an idea, a mere vision of the imagination, and transform it into a beautiful, functional structure.

Because we wanted to use the pool throughout the year, we included two heaters in the mix. One is a gas-fired heater for the spa, and the second is a heat pump for the pool. The gas-fired heater works faster but is not feasible for heating the pool because of the cost. Even the heat pump, which runs on electricity, is expensive, and as I looked at estimates of what it would cost to heat the pool year round and considered the resulting carbon footprint, I decided that we needed to consider solar power. Luckily our house sits on a large lot with a south-facing plot west of our garage that was perfect for a ground-mounted solar array. We built a 20-kilowatt-hour solar array there at the same time the pool was being constructed. With the 30% federal tax credit, we should recover the cost of the investment in about 10 years in reduced electric utility charges. I am planning to live long enough to see that day.

Now most mornings find us out in the pool swimming our laps. We generally swim for 30 minutes and monitor how we're doing on our Apple watches. Somehow, the watch will count laps (it counts one length of the pool, 25 yards, as one lap). Back in 2018, on a good day I could do 50 laps (1,250 yards) in 30 minutes. By 2021, I was down to about 42 laps (1,050 yards) in 30 minutes. Laura does about 58 laps (1,450 yards) in 30 minutes. I am winded a little more easily than I was in the old days, and I typically have to stop at each end of the pool to catch my breath. The longer I swim, the more persuasive becomes that internal voice that suggests longer pauses at the end of the pool or stopping altogether before 30 minutes. If I resist the voice long enough to complete 42 laps in 30 minutes, I feel victorious. If I do fewer laps, I tell myself, "Not bad, for a guy who has ALS."

I have learned that exercise produces brain-derived neurotrophic factor (BDNF), which is a protein in the brain that supports the survival and development of neurons and synapses. Since ALS is a disease of the nerves, I want all the BDNF I can get to protect my existing nerves and to build new ones. Every time I swim, I take comfort in knowing that I have released a good dose of BDNF in my brain.

As the weather gets colder, we start out in the spa with the water at 104 degrees and warm up with a cup of coffee before swimming our laps, and then jump in the spa again after we're done to take the chill off. On chilly mornings the steam will rise from the water, and often as Laura and I sit there in the spa letting the heat seep into our muscles and bones, sipping coffee as the sun peaks over the eastern horizon, we'll look at each other smiling at how good our lives are, at least for now.

We control the pool with a program on our computer that tells us the temperature of the pool and the spa. In the summer we don't use the heat pump, and from the ambient heat the water will regularly be in the 80s or even over 90 degrees, which is a little too warm. I like it at about 84 degrees. In the winter, the pool will cool down below 80 even though we set the heat pump to maintain 80 degrees. We'll swim until it gets down to about 67 degrees. Then we'll wait for it to warm up. Usually we don't have to wait long, because the weather and the heat pump will bring it back to above 67 unless it's an unusually long stretch of cold for Gainesville. We thought about covering the pool, but it's just too big. The cover would be huge and awkward to put on and take off.

We supplement the swimming with the "seven-minute workout," which we found in *The New York Times*. The workout is scientifically designed to provide a total body workout in an efficient amount of time. It involves a series of exercises done in 30-second segments with 10-second intervals to rest and get positioned for the next exercise. The exercises, like jumping jacks, wall sits, and pushups, require no special equipment other than a chair (we use the ledge of our spa, which is at chair height) and a pad for the floor exercises. We try to do it at least three times per week. If we're traveling and we skip a week we can feel the decrease in our strength. We extend the seven minutes a couple more minutes with some hip exercises that Laura got from a physical therapist to address a cranky hip.

I've found that if I do the seven-minute workout before the swim, I typically stop at about 20 minutes instead of doing the entire 30 minute swim.

The exercise allows me to sleep soundly at night, though I sleep soundly even when I don't exercise. Shakespeare had it right when he said that sleep "knits up the raveled sleeve of care," and the experts say that sleep is essential for healing. I grew up in the era that believed sleep was for the weak, and that it was admirable to get by on just a few hours of sleep. Politicians and corporate titans bragged about needing only four hours of sleep each night. Thankfully that attitude is changing, and more people seem to realize that sleep is essential for good health. We all feel better after a good night's sleep.

We track our sleep with an app called "AutoSleep" on our iPhones which works with our Apple watches to rate our sleep, as long as we wear our watches to bed. The app automatically senses when you've gone to sleep, apparently by lack of movement and slowing heart rate, and when you get up. It produces a report that quantifies the total sleep time as well as the amount of your sleep that was "light," "still/restful," and "deep" as well as any interruptions during the night. It has been fun to see the correlation between the data from the app and how we feel. We see that if we've not been exercising, we get less deep sleep, and we generally feel a little less rested. The app has made us more conscious of the importance of getting to bed at a consistent time and getting sufficient deep sleep.

After my voice problems, the symptom of my ALS that most troubles me is the decrease in my energy level. I am often tired. But sleep, of course, helps, and I've given up on feeling guilty about naps. If I'm tired during the day, I'll take a nap. It is wonderful to lie down in the afternoon and surrender to sleep. I do set an alarm, however, because when I haven't done so, I'll sleep for two or three hours and then find myself lying awake when I go to bed at night. So I usually set an alarm for an hour nap, and it startles me as it wakes me seemingly minutes after I lie down.

During the week, we get about eight hours of sleep during the night, and I get up when Laura has to get up to go to work. On the weekends or when Laura is traveling, I don't set an alarm and will often sleep for 10 or 11 hours. I imagine that my body is doing a lot of healing during that time.

As for avoiding stress, I've adopted a stress-free attitude toward life. I don't let the little things bother me, and I've decided they're all little things. Except possibly the ALS, and for that I focus on where I am today. Today, I'm doing pretty well. I had a swim this morning, I'm about to eat a big salad with chicken and fruit, I can still talk, and my mind is good, relatively speaking, that is.

But what gives me greatest hope, hope even beyond life or death, is my family. I am surrounded by extraordinary people whom I love and admire and who love me back. Laura is the sun around which I spin. She sustains me, protects me, and gives me hope. Paul and Maddie, those babies whom we brought into the world, have turned into magnificent young adults with fascinating careers in science and law, respectively. Paul's Susan is wonderful. I love them all and know that they love me.

Then extended family and friends, from those I see often to those I see rarely, are precious to me. Precious not because I have some deep emotional bond with them, but precious because they are my community, the people I recognize and identify as my relatives, friends, colleagues, business associates. They all give me a sense of belonging, kinship, and camaraderie. I feel good to be part of this tribe.

There's a question on one of the surveys that I have to respond to each time I visit Emily Plowman's lab, which asks: "How is your health?" The multiple-choice responses that are offered are "poor," "fair," "good," and "excellent." I always pause at this, thinking, *If I have ALS, isn't my health by definition poor?* Or is the question asking apart from your ALS, how is your health? It's like asking, "Other than that, how did you like the play, Mrs. Lincoln?" I usually compromise by responding "good," because I'm eating right, getting plenty of exercise and sleep, and avoiding stress, and I'm surrounded by people who love me.

Chapter 15:

Plasticity

Brain scientists say that our brains have the ability to create new neural networks and synapses throughout our lives. They refer to this ability as "neural plasticity." Of course, we have greater neural plasticity when we're young, but you can teach old dogs new tricks also. It's important, therefore, that we remain engaged in the world, having new experiences and learning new things, especially as we age.

With this in mind, as we became more concerned about my health in about 2014, Laura decided we should take up guitar. She had always wanted to play guitar and thought it would be a great way to have fun and stimulate our brains to create some new neural networks.

I didn't oppose the idea, but certainly wouldn't have undertaken it on my own. I imagined playing badly and feared I would be mired in frustration as my ALS prevented any progress. Furthermore, I would be starting with a disadvantage due to some finger injuries in my past. I am unable to move the joint closest to the fingertip of the index finger on my left hand as a result of slicing the tendon for that joint when I broke a wine glass while washing it in the sink. Also, I smashed the ends of the ring finger and pinky finger on my right hand when I was 12 years old by getting them caught in a farm implement my brother and I were attempting to attach to a tractor. Nevertheless, I decided I should give it a try primarily to indulge Laura's desire to do it.

Now, years later, I think I enjoy the guitar more than Laura, or, at least, I play more because I am home all day and not nearly as busy as she is. We bought two Washburn acoustic guitars, nothing fancy, but they're beautiful, each with a gleaming spruce wood top surrounded by mahogany sides and back, a dark fretboard, silver tuning pegs, and six strings running down the center over the sound hole to the bridge. I like the feel of the guitar in my hands and the feel of the strings beneath my fingers as I hold chords. And the sound. Even single notes stir something in me. In those rare moments when I'm able to put a phrase together that sounds like it should, I feel like I've accomplished something wonderful. I allow myself to enjoy every small advancement. I want to improve, but I'm careful to allow myself to fail without guilt, because the point is to have fun and build some new neural networks.

We found a teacher who embraces our approach perfectly. Randy Walker of Gainesville's Academy of Art and Music is in his 60s and grew up in Gainesville when it was a fertile ground for rock-and-roll guitarists. Tom Petty, Stephen Stills of Buffalo Springfield and Crosby, Stills & Nash, and Don Felder and Bernie Leadon of the Eagles all called Gainesville home in their formative years. Randy wears his gray hair long with varying amounts of facial hair and has a laid-back, friendly manner appropriate for an aging rocker. He is a good teacher for us, setting a path for us but never pushing us beyond what we were willing and happy to undertake. He constantly encourages us and celebrates our modest achievements in a way that demonstrates a genuine enjoyment of sharing his love for guitar with others.

When we started finger picking, Randy gave me a metal banjo pick to put on the third finger of my right hand to make up for the missing tip, and told me about various great guitar players that had missing fingers, thereby making me appreciate how minor my injuries were. It's been a long process, but thanks to plasticity, I am slowly getting my fingers to behave to pick an arpeggio pattern on the chords. I tell myself, "Not bad, for a guy who has ALS."

Although I enjoy music and was in choir in high school, I've never played an instrument. I wish I'd started earlier. I think of music as a universal human language. It communicates sadness, happiness, love, sexuality, longing, whimsy, strength, order, patriotism, humor, and more in a way that often reaches deeper into the soul than the spoken or written word. We use music to stir emotion, to strike those chords that resonate in our brains. It's no wonder that political candidates carefully select the play list for their rallies to match the message and emotional tone that they are trying to communicate.

So when I sit down to play guitar, I'm engaging my senses of touch, sight, and hearing in a disciplined way that involves both my rational thought and my emotions. In other words, my brain is getting a good workout. Because this is all relatively new to me, my brain is mapping out new neural networks to meet these new demands that have been placed upon it. When I've been spending too much time at the computer and I begin feeling lethargic, I find that I can walk to the other room and play guitar for 10 or 20 minutes and feel refreshed. And it's fun!

Chapter 16:

Legacy

I was the last of my parents' four children, and maybe being the youngest instilled in me a desire to catch up, to be good enough, to be acknowledged. I tried hard. I was competitive and studious and wanted to make a difference in the world. But my dreams of football glory ended with high school when I decided I was too small for college football. My dreams of glory on the track ended with college when I realized that no matter how hard I trained, I would never be fast enough to win the big meets or go to the Olympics. My dreams of academic glory ended when Harvard, Yale, and Stanford rejected my law school applications. Like most of us, I began to realize that you win some and you lose some and that in order to be happy, you have to be satisfied with doing your best at those things that you can control and let the rest take care of itself. Unfortunately that is harder than it sounds, but I have gotten better at it as I've gotten older.

Because of this constant inclination toward self-evaluation, as I approach the end of my life, I start to ask myself, "Have I made a difference?" What will I leave behind when I'm gone that I can feel good about?" Although there are a few things that come to mind from my career, the great accomplishment of my life is my family—Laura, Paul, and Maddie with the recent addition of Susan, Paul's wife. I am proud of each of them. The thought of grandchildren makes me smile with what I imagine is deep, evolutionary, patriarchal pride.

However, my gene mutation hovers over this familial legacy like a curse. Both Paul and Maddie received one *C9orf72* gene from Laura and one from me. That means they had a 50% chance of receiving the expansion mutation in my *C9orf72* gene. If they inherited the mutation, they would be at risk of developing ALS. Although my mom and I developed ALS in our 60s, a cousin of mine developed the disease at 35, so there is a risk of serious illness even at a young age. If either Paul or Maddie had the mutation, his or her children would also be at a 50% risk of inheriting the mutation.

Since the *C9orf72* gene was associated with ALS in 2011, scientists have been able to screen patients to determine whether they have the mutation. Having this test done, however, requires careful consideration because living knowing you have the mutation is different than living not knowing. In my experience, if you are at 50% risk of having the mutation and you haven't been tested, you are likely to assume the best case, and you will live in about the same way you would if you didn't have the mutation. Although we didn't know the gene in our family until after 2011, based on the inheritance pattern in my extended family, Laura and I suspected that there was a 50% chance that I would inherit whatever had caused my mom's ALS. Although I knew that intellectually, it didn't really affect my day-to-day thoughts until I began to be affected with changes in my voice. If I had known I had the gene, however, I suspect that news would have weighed more heavily upon me. I would have had to think harder about having children, for example.

So I advised Paul and Maddie to take their time and think hard about whether to get tested for the gene. I also said that if they were tested and they had the gene, they could start doing things now to be as healthy as possible so as to reduce the chances of developing the disease.

In late 2017, Paul was in graduate school in molecular biology at the University of Iowa where he worked in a lab that, among many other things, did genetic testing for hearing loss. He decided that he wanted to know whether he had the gene. The die was already cast, he reasoned, so the test itself wouldn't change anything. He had seen our efforts to change our diet and concluded that he might do the same if he had the gene. He didn't tell anyone and set about on a private mission to figure out how to do the test.

He later told me it was not that hard. He knew that the mutation that causes ALS was an expansion mutation in *C9orf72*, and that expansions of more than 24 CCCCGG repeats were considered pathogenic. He knew he had two versions of the *C9orf72* gene, each version referred to as an "allele," one that he inherited from Laura and one from me. I also have two alleles of the *C9orf72* gene, one with the expansion mutation from my mom and one normal one from my dad. When Paul was conceived, there was a 50% chance that he could have inherited my allele with the expansion mutation and a 50% chance that he could have inherited my normal allele.

Because the *C9orf72* gene was discovered in 2011 and has been the subject of research since then, Paul was able to order primers for the *C9orf72* gene from a lab supply company. A primer is a short fragment of DNA that is designed to be complementary to, and therefore bind to, a segment of DNA, in this case, the *C9orf72* gene. Primers are used in polymerase chain reaction, or PCR, one of the most commonly used techniques in genetics labs. Essentially, the primers identify the particular DNA sequence to be examined, and PCR makes many copies of that sequence, or amplifies the sequence, enabling examination of, or experimentation with, the sequence.

After amplifying his *C9orf72* gene through PCR, Paul planned to use capillary electrophoresis to determine the length of the alleles. Capillary electrophoresis enables fine resolution of DNA fragment sizes by running fluorescently labeled DNA fragments through a polymer using an electrical gradient. It produces a readout in graph form.

When he had mapped out his plan and determined what he needed, Paul ordered the primers, and other supplies, then had to wait for the order to arrive. When they arrived, he set everything up and pulled out a couple of hairs from his beard, making sure to collect enough for three separate replicate samples. He used a standard protocol to break down the cells in the attached hair follicles and extract their DNA. He set up and ran the PCR. He added the fluorescent probes and purified the PCR products for fragment analysis. Then, because someone else was responsible for running the capillary electrophoresis (CE) machine, Paul handed the samples to that person.

At the end of the next day, a Friday, the results were returned on a USB stick. Since Paul had not used the software for the CE machine before, he asked the CE person to show him how to use the system. After a couple of minutes, the CE person had pulled up the data and left. At first glance Paul thought that what he was seeing meant he was okay, but studied it carefully to make sure. The graph showed two distinct alleles, both of normal size, for all three samples. Paul later said to me that his focus on determining whether the experiment had worked properly precluded any sense of anxiety or excitement at the news. But as he let the data sink in, he felt very happy.

Since it was the end of the day, Paul shut down the computer, left the lab on his bike, and rode home. Once there, he called Laura and me. "I ran a test, and I think I don't have the gene," he said. We were shocked because we hadn't known that he had been doing this but hopeful that it was true. Laura began to ask him technical questions about the process he used, and as he responded, she began to cry tears of joy. Then he got in he got in his car to drive to Des Moines. At that time in 2017, Susan was at medical school in Des Moines, and Paul often drove the two hours from Iowa City to Des Moines to spend the weekend.

Then we started to think about Maddie and how to tell her that Paul did not have the gene. We were concerned that if Paul told Maddie this news, Maddie would begin to think about whether she should be tested, thereby adding stress to Maddie's already stressful life as a first-year law student at Yale. It was the fall of 2017, and Maddie was preparing for end-of-semester finals. We asked Paul to wait to tell Maddie until we were all together at home in St. Paul for Christmas.

Paul's next call was to Susan. He and Susan had talked previously about the ALS in our family and the risk that Paul might carry the gene. Because he didn't regularly think about it, Paul had assumed that Susan was similarly unconcerned with the risk, but when he told her that he did not have the gene she began to cry and was thrilled at the good news.

A few months later, Paul, Maddie, Laura, and I were at 2116 Carter Avenue in St. Paul, enjoying a glass of wine after dinner, when we told Maddie that Paul had tested himself for the gene and did not have it. Maddie was surprised that Paul did the test without talking with her about it first. They are close, and Maddie thought that since they shared this biological risk factor, Paul would have talked with her before going ahead with the test. We all talked about the pros and cons of getting tested or not, and concluded that there was no rush.

For the next year, Maddie, in true lawyerly fashion, went over the arguments for and against getting tested. One of Maddie's classmates at Yale, a new father of about 30, had recently been diagnosed with ALS and was already having trouble walking. The news had spread quickly at the law school. When the subject of the classmate's ALS came up during a conversation with a clinical professor who knew the classmate, Maddie suddenly found herself in tears, and, embarrassed, explained that she had ALS in her family too.

Maddie recalled the *Gleason* movie, and how rapidly ALS could progress and the burdens it placed on the spouse/caregiver. She wondered whether she would be viewed as flawed by potential spouses if she tested positive for the gene, or perhaps, even if she were not tested but simply at risk for the gene. She eventually decided it would be better to know whether she carried the mutation. Laura and Paul discussed the details about running a test, and Paul agreed to run the test.

We waited until the winter break of Maddie's third year when she was visiting us in Florida for a couple of weeks during January 2018, when the pressure of law school would have abated for a time. Paul said he would run the test sometime during the visit. Laura took blood samples from herself, Maddie, and me, extracted the DNA, and sent the samples to Paul. Paul added his own DNA. He then ran the test on all the samples.

Soon after, as Laura was driving home from work, she saw that Paul was calling. She pulled over and stopped the car so that she could have her full attention on the call. Paul said, "I think we have good news." Laura began to cry. He sent a picture of the graph from the CE machine, and Laura examined it and agreed with his conclusion that Maddie did not have the gene. It was clear from the graph that each of Paul and Maddie had inherited my one normal allele, but each had inherited a different normal allele from Laura. Laura came home and told Maddie and me. We hugged and celebrated the good news.

For me, the dark clouds of my fears had disappeared, and the sun shone brightly. My legacy was secure. I had not passed this mutation to future generations. I can live and die without regret as long as Paul and Maddie and my future grandchildren do not have this gene. I've lived a good life. I've made a difference in the world.

Maddie had mixed feelings on hearing the news that she did not have the gene. She was happy but said that part of her was sad to abandon me as the only one in the family with the gene. I love her for that thought. Similarly, I feel a bond with my long-departed mother that arises from the fact that we share this gene and this dreadful disease. I am not alone; she went before me and gives me strength that I can do this too.

Chapter 17:

Vickie

For my 60th birthday party in 2018, I asked each of my siblings to share a memory from our youth and to write it out in advance and be prepared to read it at the party. Vickie read a story from about 1963 when I was five and she was nine and we had just moved from Grand Forks, North Dakota, to a lovely place by the Red Lake River five miles outside of Crookston, Minnesota. Dad built a house there, slowly, on the weekends and evenings, hammering every nail and laying every brick himself. Across the road and up a hill, our neighbors had a farm.

Shortly after we had moved in, Vickie became aware, somehow, that there was a birthday party taking place for one of the neighbor kids and, significantly, that there were cupcakes involved. She was hungry and decided that we had to get some of those cupcakes, so she told me to go over to the neighbors and ask if we could have some. I have a vague memory of walking up the hill with some trepidation only to be received warmly by an older woman whom I came to know as Martha, the grandmother who lived in a trailer home beside the farmhouse. She gave me four cupcakes, and I carried them back down the hill and across the road to my waiting siblings. We were all excited at the success of our caper. Then Vickie took the cupcakes to bring them to the house and dropped them all into the dirt of our construction site yard. They were irretrievable.

The story illustrates a couple important things about Vickie. First, she was sufficiently older than I was to have some authority over me when I was young, and that separated us as we grew up. Although we lived in the same family, we grew up in two different worlds and lived in two different worlds as adults. Vickie did not go to college, became involved in the Baptist church as a young woman, married a man from the church, and settled into domestic life. Second, she had a sweet tooth and always loved baked goods. As a result, she was a master in the kitchen and created a loving home for her family. She was the nicest person in the world, full of kind words and actions, and unable to speak harshly of anyone.

Shortly before Christmas in 2019, I got a call from Vickie. She was in her car. She started as she always does: "Hi, Rob. How are you?" Her voice had a nervous tone.

"I'm fine, Vickie. I'm here at my computer going through emails. It's a beautiful day down here. How are you?"

"Well, Rob, I wanted to tell you about some weakness I've been having."

My heart sank. I thought of her Christmas card: Vickie and her husband, Steve, lined up with their three daughters, their husbands, and all the grandkids, all in front of an old log barn. Vickie has raised three lovely daughters, Becky, Bridget, and Sarah. Becky has six kids. Bridget has three, and Sarah has two. If Vickie had the ALS gene, there would be a 50% chance that each of Becky, Bridget, and Sarah would have it, and all 11 grandkids would be at risk. I felt suddenly heavier, as if some energy had left me.

She continued, "For several months now, I've been having trouble carrying things up from the basement, or, for example, Steve was putting up some panels in the basement ceiling, and wanted me to get up on the ladder and hold the panel in place while he did something, and I just couldn't hold it over my head for any time at all."

I asked, "How long has this been going on?"

"I really started to notice it in the fall, but it may have been going on all summer. I just don't have the energy I used to have. I chalked it up to old age at first. When I had my 65th birthday in September, I thought, wow, I really feel older. You know, I used to have lots of energy and could go all day working in the kitchen, gardening, doing laundry, but now I just want to sit down with a cup of coffee. So I think I should get tested for the gene. Does Laura still have my blood sample?"

Laura had gotten blood samples from all my siblings a year earlier.

I said I would ask Laura. I asked if she was having fasiculations, the involuntary twitching, quivering muscles that indicate nerve damage. She said no, but she has experienced itching in her legs, as if her skin was irritated, although the skin looked perfectly normal. I asked about her voice and swallowing. They were fine. Handwriting, maybe a little slow. Eating, normal.

When we ended the call, I called Laura right away. She had the sample at the lab but said the lab would shortly be closing for the holidays. She would have to check on whether she had the appropriate written consent to use the sample in research. If so, she said she would have someone screen the sample for the gene after New Year's Day. Talking to Laura restored my equilibrium. We would get Vickie tested for the gene, and if she had it, we would get her on my program, taking metformin, exercising, sleeping, and eating well. We would do the best we could do.

On January 2, 2020, Laura got the news that Vickie did indeed have an expansion mutation in the *C9orf72* gene. Based on Vickie's symptoms, the news was not a surprise. We talked about whether Vickie should participate in the clinical trial for metformin that Laura and others at UF were organizing. Given the encouraging results in the preclinical lab work with my mouse cousins, Laura had gotten support from UF to run a small clinical trial to determine whether metformin was safe, tolerable, and a potentially viable therapeutic treatment for *C9orf72* ALS. The FDA had granted a waiver from their normal review process because metformin was already FDA approved and well known as a safe drug. The trial, which would be conducted at the University of Florida, was just beginning to enroll patients and had an estimated total enrollment of 18. Participants would have to endure several lumbar punctures to obtain cerebrospinal fluid to measure for RAN proteins and determine whether taking metformin reduced the level of RAN proteins. All participants would receive the drug, and there would be no placebo group.

Laura and I called Vickie and explained that she had the C9 gene and discussed the trial. She was quiet and serious on the call, and I knew from my own experience that this was hard for her to hear. We did our best to lift her up. We explained that I was doing well on the metformin and that based on the data from the preclinical work in mice we believed that it could help her too. We encouraged her to participate, saying we would love to see her, that she could stay with us and have a little vacation from the Minnesota winter. We explained that, alternatively, she could ask her doctor to prescribe metformin off label, as I had, and start taking it right away even if she didn't participate in the trial. I added that if she participated in the trial, she would have the benefit of seeing all the great doctors at UF and getting answers to any questions. She recognized that by participating in the trial she could add something to the fight against this disease and wanted to participate for that reason.

Vickie's screening visit for the metformin trial was scheduled for late January. Cal agreed to travel to Gainesville to be with Vickie for the first visit. Cal and her husband Wes live in Jefferson City, Tennessee, about 40 minutes by car outside of Knoxville. Cal was able to fly from Knoxville to Gainesville with a connection in Charlotte, North Carolina. Cal arrived in Gainesville first, and I picked her up and brought her to our house. Laura was at work.

Vickie doesn't travel by air much, and so it was a big deal for her to travel from Climax, Minnesota, to Gainesville, Florida. Her daughters live about an hour north of Minneapolis, and so Vickie drove about five hours from Climax to Bridget's house and then stayed overnight before taking the flight from Minneapolis. Bridget and Sarah drove her to the airport in the morning, parked in the ramp, and walked into the airport ticketing area with Vickie to make sure she knew what the right gate was and got into the right security line. Vickie raised those girls to be as generous, kind, and helpful as she is.

The Delta flight from Minneapolis connected in Atlanta, and we had coached Vickie on how to find her connecting flight to Gainesville in the Atlanta airport. We were relieved when she texted that she had found the correct gate and had boarded the flight to Gainesville. Cal and I went back to the airport to pick up Vickie. I noticed right away a change in Vickie's posture as she walked. She carried her torso differently, but I didn't say anything, and welcomed her enthusiastically.

We had time for lunch before Vickie's first appointment, and so I made a salad when we got back to the house. I made a typical lunchtime salad for each of the three of us that consists of a dinner plate full of greens (red leaf lettuce, spring mix, romaine, spinach, kale, or arugula, depending on what was in the fridge), red onion, carrots, cucumber, fruit (one or more of pears, blueberries, raspberries, or apples) and some kind of protein, probably leftover farmers' market chicken from dinner the night before. The olive oil and salt were ready on the table. Vickie took photos of the salad to send to her daughters.

Although I do not regularly pray before meals, I knew that my sisters did, and so I was not surprised when, shortly after we sat down, Cal said, "Let's pray," and held out her hand. Vickie followed her lead and reached out from the other side. I grasped both their hands, and Cal began, "Dear Lord, we thank you for this opportunity to be together and for this wonderful food and for the opportunity for Vickie to participate in this trial. Please strengthen Vickie and Bert and cure them of this disease. Guide the doctors in their work and make them tools of your healing grace. Fill us all with your presence and guide us all as we live our lives in your service. In Jesus' name we pray. Amen."

I eat quite slowly, and when I'm talking, it goes slower still, and so we talked and ate for over an hour. I'm sure my sisters looked at this as their golden opportunity to win me for Christ, and so the conversation turned to religion. Bible verses were quoted.

We were like America, divided along educational, economic, and religious fault lines, but tied together by a shared history. I decided that I had a responsibility to reach across the divide with love; these were my sisters. So I laid my cards on the table.

"I used to be religious," I said. "I went to church and prayed. I was sincere. But as the years went by, doubts crept in. My prayers were all one-sided conversations, and there was never any answer. And although I loved the music, the ritual, and the acknowledgment that there is more to our existence than what we see day to day, there is so much that religion gets wrong. It divides us. People kill in the name of religion. Leaders use religion to pursue status or worse: to exploit their followers. Think of all the priests who turned out to be pedophiles."

I had seen an article that morning in the *Gainesville Sun* about a house for sale in a little town called Micanopy, about 40 miles south of Gainesville. The house had been owned by a Christian cult called the House of Prayer led by what must have been a charismatic woman. She was in jail awaiting trial for the murder of a two-year-old. She had been arrested earlier on charges of child abuse for bathing a 12-year-old child with bleach and causing burns. The article said that she controlled her flock through "fear and charisma," and that "Residents—adults and children—were routinely beaten or kept confined in small areas, and food was withheld."

The paper was on the table, and I handed it to Cal, saying, "Take a look at this story about a Christian group in Micanopy. They prayed and believed in Christ."

Cal read for a moment and said, "Ughh, I can't read this. Our church is not like this. This would never happen in our church. There is evil in the world, and that was a terrible woman, but you can't condemn the Christian church because of one person's actions."

"I agree," I responded, "but it seems to me that people are people, some good and some bad, and it doesn't matter whether they claim some religion or not. Religion does not change people; it's just a story they adopt."

"It changed me," Cal said. "After Wes and I were married and we moved to South Dakota, I just felt empty. I was fine, going about things day to day, but I felt like there should be more purpose, more something. The pastor of this little Baptist church showed up at our door with a pamphlet, and I originally thought 'No way, we'll never go there.' But then time went by, and I saw him in town again, and he said 'Hoping to see you soon.' Pretty soon I was thinking about it all the time. God was leading me to that church. So one Sunday, I convinced Wes to take a drive up there. It felt like going home. I accepted Jesus Christ as my savior and was truly born again. I was made new. I was absolutely a new person. That emptiness I felt was filled. I began to read the Bible and pray more. I really feel that was a turning point in my life. And so that's what I want for you, for everyone, really. And I know that the Bible is the one way."

Vickie was nodding her head sincerely.

"I'm happy that you have found something that works for you, but I hope you can understand my doubts. I have to weigh what I have experienced and learned in my 60 years on this planet against what you and the Bible are telling me. Much of what the Bible says is consistent with human nature, how humans see the world in literature and stories, and inconsistent with my experience in the world and what I have learned about science. Generally, virgins don't have babies, people don't walk on water, and people don't rise from the dead. Those kinds of things happen all the time in stories that people make up. So I just don't know what lies beyond this world. But I'm okay leaving it as a mystery."

Time was running short, so we continued the conversation as we cleaned up the dishes and got ready to leave for Vickie's appointment. I felt good about how things had been left. They knew where I stood, and although this made me an "outsider" in their eyes, we were still siblings that had grown up together; they loved me as their brother, and I loved them, my sisters.

We drove to the Norman Fixel Institute for Neurological Diseases, a handsome new building south of the campus of the University of Florida, checked in, and waited in a beautiful waiting room with a wall of windows overlooking a wooded area behind the building. Dr. Jim Wymer's assistant, Jennifer, appeared and escorted us back to an examination room where all three of us—Cal, Vickie, and me—sat while Jennifer gathered information from Vickie, weighed her, and took her blood pressure and heart rate—in other words, got her "vitals." This was the "screening" part of the visit. Eventually, Dr. Subramony appeared.

Laura has known "Sub" for years from the neurological community, even before Sub came to the University of Florida in 2009. When Laura and I moved to Gainesville in 2010, he was already a part of the community of neurological doctors and researchers that Laura was joining and was especially warm and welcoming to us. We enjoyed several social occasions at his home and invited him to our house whenever we had a gathering of UF people. He is our friend. Sub got his medical degree in India, his native country, and then came to the

U.S. for training in neurology at the Cleveland Clinic and stayed. About 70 years old, he has a warm and friendly manner befitting his status as a doting grandfather to two boys in New York.

Sub chatted with Vickie for a bit and then did the push-pull neuro exam that I had been through so many times, watched Vickie walk a straight line, and poked her in various places, asking, "Can you feel this?" Then he began to fiddle with an EMG machine, complaining that this was a different machine than he was accustomed to. Eventually he saw the response he wanted and began to insert needles in Vickie as she lay on an examining table, listening to the feedback sound as Vickie alternatively moved and relaxed the muscles being examined.

In the end, he told us that Vickie definitely had some weakness on her left side. He said that neurological damage had probably been taking place for years based on what appeared to be compensatory growth in other nerves to take the place of the damaged nerves. Vickie was surprised at this and said she had only really noticed any symptoms within the last year.

Then Dr. Subramony left, and shortly after that Dr. Wymer arrived. Dr. Jim Wymer came to UF after Laura but along with Sub and others, he is part of the neurology gang that Laura works with. He is also a friend and is one of my doctors. He was involved in the planning and organization of the metformin clinical trial. He is an energetic man about 60, with glasses and well-trimmed brown hair that in his younger days likely covered his entire head. He has a happy disposition and is always looking for a reason to laugh. He quickly put Vickie at ease and explained that he was going to perform a lumbar puncture to remove some cerebrospinal fluid to get a baseline sample before Vickie started taking the metformin.

Cal and I were excused from the room for the lumbar puncture. As I was leaving, I said to Vickie, "Don't worry, Vickie, it's a piece of cake."

When Jennifer came and retrieved us from the waiting room, Vickie confirmed that it wasn't bad. She was instructed to lie flat for an hour after the LP, and we chatted with Jennifer about the trial, kids, and travel to pass the time.

After leaving the Fixel Institute, we were instructed to drive a short distance to the UF blood draw laboratory at Rocky Point. Despite the enticing name, the facility was a nondescript building with a parking lot. Maybe the Rocky Point was somewhere nearby but out of sight. There Vickie had some blood draws completed, including one to confirm that she had the *C9orf72* mutation, and then we were free. In the car, Vickie said, "You know, no one ever said that I have ALS."

I thought for a moment about my own experience coming to terms with the diagnosis, and responded,

"I'm sorry to say it, but yes, you have ALS. We're in this together, Vickie. You have weakness, and you have the *C9orf72* gene. They just want a certified lab to confirm that. You're in the trial, and they only take ALS patients in the trial. So you have to get used to that fact, but stay positive. We need to do everything we can to stay healthy while the science closes in on a cure. This metformin is a first step. They'll know more in the future. But we need to stay around to get the benefits of that progress, and that means we have to exercise, eat healthy, sleep well, and avoid stress. I don't do stress anymore."

I told Cal and Vickie about the sleep app on the iPhone, "Autosleep," that records my sleep with my Apple watch.

"During deep sleep, your body is repairing all kinds of stuff," I said. "You should get as much deep sleep as you can to help your brain and nerves repair themselves."

They were interested, and since we had several hours, we went shopping for Apple watches. We found a good deal on earlier models at a Sam's Club, and Cal and Vickie each bought one.

In the afternoon, we went to Emily Plowman's lab for a swallow study where I was greeted like an old friend, and I introduced Cal and Vickie to the group. There Vickie sat in the chair where I had sat many times, swallowing barium substances of various consistencies while X-ray machines took videos of her swallowing each one. Her swallowing was fine, as we expected from the lack of any bulbar symptoms. Then, at 5:00 p.m., we went to an MRI lab, where they put Vickie in the MRI tube and took pictures of her brain to look for any abnormalities.

Armed with a new Apple watch and a prescription for metformin, Vickie departed the next morning on a flight for home. Cal's flight was an hour or so later.

A few days later I received an email from my brother, Mike. It read:

> Bert,
> Thanks for all you and Laura are doing for Vic. It was great to talk with you last night and it was clear she is drawing a great deal of strength from you all. As you know, this thing is potentially very scary and having you there to guide her is a big comfort.
>
> That brings me to my second point. I have always thought that knowing whether I had the gene marker or not was not helpful because there was little or nothing that could be done about it. However, that is changing fast. It seems to me that knowing might be useful. While I don't have any sense that I have any weakness or other symptoms I would like to ask if Laura still has

my blood, if she could do the analysis to see if it contains the gene marker. If she doesn't have any blood we should think about how to make that happen. Of course, we will pay for any costs.
Let me know what you think.

Thanks again for taking care of our sisters!
Mike

I was surprised at this because Mike had always said that he didn't want to know, but I understood his rationale. Things were changing. There were things that could be done now to possibly avoid the decline that I'm convinced begins long before symptoms are noticeable. Laura had Mike's DNA stored in the lab and instructed one of her lab members to run the analysis.

During the one or two days while we waited for results, I thought of Mike and Mary and their daughters Ruth and Emma. Both were married. Emma had given birth to Mike and Mary's first grandchild, a beautiful child named Ella Ruth, in late 2018. So Mike wasn't the only one with a stake in the results of the test.

Laura and I called Mike with the results.

Laura didn't waste any time and said, "Mike, I have some good news for you. You don't have the gene. They were careful in running the tests, did it multiple times, and it's clear—you're okay."

"That's great to hear," Mike said. "Thanks for running the test." Reticence runs like a deep, glacial fjord in Mike's soul. I knew from his tone of voice and the word "great" that he was excited and pleased. But I didn't expect him to say much more.

"You should celebrate," I said. "Sit down with Mary tonight and have a glass of wine and celebrate your good fortune. This is a big deal."

"I will," he said.

"All right, We'll let you go so you can tell your family," I said, signing off.

Vickie's visit had also focused the attention of Vickie's three girls—Becky, Bridget, and Sarah—on their mom and her symptoms and also the possibility that they might have the *C9orf72* mutation and have passed it along to their 11 kids, including Becky's six, Bridget's three, and Sarah's two. They must have talked about it among themselves, because after Vickie's screening visit all three contacted Laura and wanted to be tested. Laura is not a medical doctor and can't give medical advice, but she and her lab do the type of work involved in genetic testing regularly for research purposes. Becky, Bridget, and Sarah signed a form stating that they consented to be involved in ALS research. Then Laura had people in the lab run the tests.

We were concerned because statistically, the likelihood that all three would have avoided the mutation is ½ x ½ x ½ = 0.125. In other words, there was a 12.5% chance that all three of the girls would be free of the mutation. If we included Paul and Maddie, whom we already knew did not have the mutation, then the chance that all of the cousins would not have the mutation would be ½ x ½ x ½ x ½ x ½ = .03125, or 3.125%. Those were not good odds.

When Laura got the news, she called each of Becky, Bridget, and Sarah separately to deliver the report of what they had found. Then Becky and Sarah called Laura back and wanted to add Bridget and then conference in Vickie. Bridget was on another call and couldn't come to the phone, but assuming she would be available in a moment, Becky added Vickie to the call.

Becky said, "Mom, this is Becky, and Sarah and Laura and Bert are on the call."

We all said, "Hi, Vickie" or 'Hi, Mom."

Becky said, "We're trying to add Bridget to the call because we all wanted to be together to tell you some news."

Vickie said, "Some news? Oh, now I'm really curious."

Becky said, "Hold on for a moment while I try to add Bridget; she was on another call." She then called

Bridget, who picked up, but said she was on a work call that she couldn't leave and that she didn't know how long it would go.

Becky joined the call with Vickie, and said, "Mom, I'm sorry, but Bridget isn't available right now. We wanted to be all together to tell you the results of the gene test that Laura did, but we'll have to call you back."

Vickie said, "Tell me! You had the gene test! All three of you! You can't leave me hanging like this."

Becky said, "Well, I will say that it's good news, and you don't have to worry, but we'll get Bridget on the line and call you back."

Vickie said, "Good news!? What does that mean?"

Becky said, "We'll give you the details when we call back, okay? We'll call back as soon as we can. Love you."

Vickie said, disappointedly, "Okay, love you too."

When Becky, Bridget, Sarah, and Laura called Vickie back, I wasn't on the call, but I know that they told Vickie that none of Becky, Bridget, or Sarah have the *C9orf72* mutation! And that meant that none of Haakon, Marit, Thor, Syneva, Aksel, Freya, Scarlett, Anders, Violet, Lilly, or Gunnar have the *C9orf72* mutation! And neither Paul nor Maddie have the mutation! It was a happy day in our family. The 3.125% probability outcome had been realized. Some might call it a miracle.

Knowing the relief that I felt when I learned that Paul and Maddie did not have the gene, I imagined that Vickie had even a greater relief because she is a kinder soul and has a bigger family to worry about.

Vickie returned to Gainesville at the end of February for her second visit. Cal didn't join her for this visit, so I brought Vickie to her appointments. We went to the Fixel Institue for an exam and a lumbar puncture, and Rocky Point for blood draws, but there was no swallow study or MRI on this visit.

There was increasing concern about the COVID-19 virus even at the time of that visit. Soon the virus spreading through the country was a more immediate threat to Vickie's and my health than the mutations in our cells.

Chapter 18:

COVID-19

In February 2020, I made reservations at a hotel in Key West to celebrate Laura's 60th birthday on May 15, 2020. I booked flights to Key West for Susan who would be traveling from Minneapolis, Paul traveling from Philly, Maddie traveling from New York, and Laura and I traveling from Gainesville. We planned to spend a long weekend scuba diving, eating well, and generally enjoying ourselves. Even in February, I was wondering if the trip would work out given the increasing concern about the COVID-19 virus, but by March 17, I was sending around emails canceling our trip. Multiply that act by millions, and it's not surprising how the virus has affected the economy, especially the airlines, lodging, and tourism industries.

The COVID-19 virus first hit the Northeast hard. Maddie was living in Manhattan and clerking for a judge in New Jersey as the infections spiked in New York. The court soon closed its offices and had everyone working from home, and we encouraged Maddie to come to Florida, arguing that she could just as easily work from here and reduce her risk of getting the virus. Laura was so relentless with her pleas for Maddie to leave New York that Maddie eventually insisted that we stop talking about it while she considered her options. After New York hospitals began to get more crowded and there were fears about adequacy of personal protective equipment and ventilators, Maddie finally agreed to leave New York and come to Florida.

Maddie and her cat, Evie, arrived at our house on April 12, Easter Sunday and stayed until May 16, the day after Laura's 60th birthday. At that time Evie was just beginning to accept Laura and me as cohabitants of the house, but Maddie's tolerance for living with Mom and Dad had been exhausted. She wanted to get back to the city, where the virus control efforts seemed to be yielding results, and to her independent life.

To celebrate Laura's birthday, Maddie and I made Peking duck and a salad, opened a bottle of wine, and sat down to dinner with Paul and Susan joining us via the laptop on Zoom. The duck was perfect, the skin crispy with enticing aromas of cinnamon, nutmeg, cloves, and cardamom. Maddie had made little flour pancakes into which we folded slices of duck, green onions, and a sweet/sour sauce. The Key West hotel could not have done better! Maddie's impending departure added a bittersweet poignancy to the evening, and it was a lovely 60th birthday celebration for Laura.

After Maddie drove off for her 16-hour journey back to Manhattan, Laura and I walked back into the house heavy-hearted. Laura shed a few tears. We were grateful for the visit. Because of the virus, Maddie had come and stayed for over a month! She would never have done that if things were normal; she might have stayed for a long weekend, or we might have traveled together for a few days. But a month-long visit allows you to develop rhythms, to enjoy those casual encounters in the kitchen refilling a coffee cup or watching a Netflix show together. It seemed to me a throwback to an earlier time, when travel was harder and guests would plan to stay for a longer time because the trip took so much longer that one simply did not move quickly or easily. We love that girl, and it is always hard to see her go. We want to hold her close but know that we must let her go. We are the past and she is the future. For a little while, we are here together, and it is hard to let go of that time together.

The Centers for Disease Control and Prevention (CDC) says that people with "underlying conditions" are more at risk for the virus. Included in the CDC list of underlying conditions is "neurologic conditions such as dementia." Even though I don't have dementia, I assume I might have a harder-than-average time with the virus. Some reports have said that people who get the virus sleep all the time. I already sleep a lot and generally have low energy, and so if I got the virus who knows what might happen.

The virus added another level of risk to my situation. While I didn't feel any anxiety about this (it's out of my control), we did everything we could to avoid the virus. Laura's training in microbiology has given her a healthy appreciation for the steps necessary to avoid the risk of potential exposure. We stayed home most of the time with occasional forays out for curbside pickup at Northwest Seafood or to just drive around to counter cabin fever.

We had our groceries delivered by a young woman named Jenn who works for a service called SHIPT. Laura liked the care Jenn took with her order, texting with questions when she was shopping to make sure she got the right stuff, and Laura tipped her well. Jenn worked in early childhood education before the shutdown, which caused her to switch to SHIPT. But when her school opened up again, Jenn decided to continue working for SHIPT and not go back to the school because she was making four times her previous salary and she enjoyed her customers.

When the groceries arrived, Jenn unloaded them into a four-wheeled cart that we left at the end of our front walk (while we often chatted with her while standing at our front door at least 30 feet away), and then we took the cart to our outdoor counter and sink by the pool to disinfect everything before bringing it into the house. We removed produce from the new plastic bags from the store, rinsed it with water, and placed it in plastic bags that had been under the counter for at least a week from past trips. We sprayed bottles, cans, and boxes with a bleach solution and then rinsed and dried them. Then we brought everything in the house and put it away.

Although as of July 2020 the University of Florida had opened up again with restrictions following its initial shutdown, each department had rotation schedules to permit social distancing, and everyone was required to wear masks. Laura worked from home as much as she could, which meant she went into the lab infrequently. She is hypervigilant about avoiding exposure to the virus and so raises my level of attentiveness from its normal inadequate level.

We all needed to rely more heavily on individual measures of protection because our government failed so miserably to protect us with conventional public health efforts. President Biden's election in 2020 brought some sanity to the federal government, but former President Trump and Florida Governor Ron DeSantis operated from the same cynical, emperor-has-no-clothes playbook. They bragged about their too-little, too-late steps to contain the virus and told the cameras that all's well while the virus was raging throughout the country.

The federal and state governments could have done so much better. With effective leadership, people could have come to believe that mask wearing was patriotic, that it was their duty to wear a mask to protect their fellow citizens. If everyone wore a mask and stayed at least six feet apart, there could have been fewer cases of the virus and fewer deaths. But from the beginning, President Trump downplayed concerns about the virus, apparently concerned that bad news could hurt his chances for reelection. He suggested that the whole thing was a hoax cooked up by the Democrats continuing his campaign of divisiveness. He said in the early days, "One day, it's like a miracle, the virus will just disappear," and refused to wear a mask.

Governor DeSantis, a loyal Trump sycophant, similarly ignored the experts and emphasized economic concerns above health. As a result, many people in the public don't believe the virus is a serious matter and don't wear masks. Young people, especially, have continued to socialize in groups and at parties and, according to the news reports, are the reason for most of the new cases in Florida. They will inevitably spread it to the elderly and those with preexisting conditions, and the number of deaths will continue to rise. As I write this the number of deaths from COVID in the U.S. exceeds 600,000. It is a loss of life similar to the Civil War.

The bright spot in this crisis was how rapidly the pharmaceutical industry produced vaccines. As the vaccination rates increased, the COVID infections decreased.

We should have been better prepared. The World Health Organization (WHO), the CDC, and anyone who pays attention to public health, like Bill Gates, had been warning of a pandemic for years. During the Obama administration, in response to the 2014 Ebola epidemic, the National Security Council headed by Susan Rice established the Directorate for Global Health Security and Biodefense. That sounds like a prudent move from where we are today. In March 2020, however, the Trump administration was accused of dismantling the office, leaving the country less prepared for pandemics. The Trump response was that the office was folded into another directorate to streamline the organization and that no functionality was lost. Regardless of what happened, the federal government did not even attempt to roll out a national plan for how to deal with the pandemic.

When Trump was first elected, I knew it would be bad, but I expected that checks and balances would provide some constraints and that the bureaucracy would continue to operate largely as before. I was wrong. Under Trump's leadership, the United States pulled out of the WHO, and President Trump and his allies regularly undermined the experts at the CDC. The bureaucracy needs leadership, and under Trump all it got was toxic disinformation and conflict.

In the midst of all the COVID-19 chaos, a murder in Minneapolis ignited a firestorm of protest across the nation and internationally. On May 25, 2020, Minneapolis police were called by a convenience store clerk who reported that George Floyd, a black man, had bought cigarettes using a counterfeit $20 bill. When the police arrived at the scene, they found Mr. Floyd sitting in a car with friends and arrested him, handcuffing his wrists behind his back. When they attempted to put Mr. Floyd in the police car, he resisted, complaining of claustrophobia and fear. The police wrestled Mr. Floyd to the ground. Noticing the altercation, a seventeen-year-old girl on the sidewalk nearby began recording the incident on her cell phone. The cell phone video, which in the days following was viewed by millions, showed Mr. Floyd on his belly on the pavement, face turned toward the camera, hands cuffed behind his back, with three police officers on top of him. The officer most clearly shown in the video, who is white, is Derek Chauvin. Officer Chauvin had his knee on Mr. Floyd's neck.

Officer Chauvin kept his knee on Mr. Floyd's neck for at least eight minutes and 15 seconds. Mr. Floyd is heard in the video saying that he can't breathe and is calling for his mother. Onlookers pleaded for the officers to get off him. Mr. Floyd became unconscious, and officers could not find a pulse. Paramedics arrived. Still, Officer Chauvin had his knee on Mr. Floyd's neck. By the time Officer Chauvin took his knee off of Mr. Floyd's neck, seventeen minutes after he had first arrived at the scene, Mr. Floyd was showing no signs of life. He was later confirmed dead.

The video burst into the public consciousness as the newest and most powerful example in a long line of police killings or mistreatment of black men. Protests against police violence began immediately in Minneapolis and other major cities and continued for weeks. The news covered the story for weeks also, including a memorial service in Minnesota and the funeral in Texas. I was mesmerized by the coverage. It seemed a significant event in the 400-year struggle of black people to shake off their history of slavery.

My illness and my advancing age cause me to take a longer view of things, and so I've been thinking about how the current unrest, including COVID-19, the reaction to the George Floyd murder, and the political chaos caused by Trump including the storming of the Capitol on January 6, 2021, compares to other times in my life. I was born in 1958, and so I was only 10 when Martin Luther King, Jr., and Bobby Kennedy were killed and the Vietnam War was in the news. I remember talk of the draft and of young adults worried about the lottery and whether their draft number would be called. Even at that age, I was aware of draftees going to Canada, which because we lived in Northern Minnesota, was only 90 miles away. Then Minnesota's happy warrior Humphrey lost to Nixon, the Vietnam War dragged on, and Watergate led to Nixon's resignation. I came of age in those years, and what I took from them was a healthy mistrust of government. The government would send you off to die in unnecessary wars. Presidents could be corrupt and break the law to secure their reelection. Because the Republicans seemed to deserve most of the blame, I voted for Jimmy Carter in 1976, the first election in which I was eligible to vote. Despite my mistrust, I believed then, and continue to believe, that government can work with the right leadership and that the United States is exceptional because of its form of government.

Maybe we are in the midst of an upheaval similar to the late '60s and early '70s. The civil unrest and mistrust of government seems similar. As then, I blame the current mess on the Republican Party. Their embrace of Trump and his xenophobic and racist pandering to our uglier instincts is one of the worst episodes in our political history, similar to the McCarthy era. Joe Biden's victory in November 2020 was an indication a majority of people have seen enough of Trump to understand that he had nothing to offer us. President Biden has already restored some integrity, decorum, civility, and order to our government. But there remains a frightening segment of the population that consumes the divisive rhetoric from Fox News, embraces wild conspiracy theories like Q-Anon, and looks to Trump as their leader. Rather than showing real leadership in the public interest, the Republican Party shamelessly panders to this fearful, largely white, rural, and uneducated group. As in the civil rights era, continued progress toward justice and fairness in civil society depends on rational people of good will rising up to thwart the efforts of those who refuse to change and insist on living in the past.

Regarding the racial issue, do all the protests, media coverage, and resulting discussion following Mr. Floyd's murder reflect a durable awakening to the injustices that still permeate our society for people of color? I hope so. Robin DiAngelo, an antiracist speaker and writer, uses the term "White Fragility" as a label for those white people who insist that they are not racist and refuse to see that they have essentially been marinated in white privilege for their entire lives and so have advantages, including psychological foundations of self-worth, that are denied to people of color.

I suspect she's right that the benefits of being white are so ubiquitous in the Western world that it's difficult to even identify them. In addition to the obvious ones, like opportunities for economic success or leadership roles, antiracism ideology as espoused by Ms. DiAngelo and Glenn E. Singleton, another well-known antiracism trainer, includes attributes of white culture such as the "ideology of individualism," which holds up what they would say is a myth of meritocracy as a neutral means through which hard work and talent are justly rewarded. Other aspects of white culture that tend to marginalize people of color, according to the antiracism ideology, are valuing the written communication over other forms; scientific, linear thinking; and deferred gratification.

All this reminds me of Edwin Moses, one of the heroes of my youth. Because I ran on the track team and paid attention to track and field in the years leading up to the 1976 Olympics in Montreal, my brother and I and some friends decided to go to the Montreal Olympics. Since we lived in Northern Minnesota, we drove there, and since we didn't have a lot of money, we stayed at a campground. We were in Olympic Stadium to see 20-year-old Mr. Moses win the 400-meter hurdles with a new world record, eight meters ahead of U.S. teammate Mike Shine, a white man.

I also ran the hurdles and so admired Mr. Moses as casual golfers admire Tiger Woods, with great respect for talent and devotion to purpose that far exceeds our own. Not only did Mr. Moses have physical abilities that were unusual even among elite athletes—he could run the 400-meter hurdles faster than the average college athlete can run 400 meters without hurdles—but his work ethic was legendary. Surprisingly, however, athletics were not a priority for Mr. Moses as a high schooler. His parents were educators, and young Mr. Moses focused on education, landing an academic scholarship to Morehouse College, where he majored in physics and engineering.

The United States boycotted the 1980 Olympics, in the midst of an unbelievable winning streak by Mr. Moses. From 1977 to 1987, he won 122 consecutive races, including 107 consecutive finals. He won the gold medal at the 1984 Los Angeles Olympics and won the bronze at the 1988 Seoul Olympics.

Would the antiracist ideology label the Olympic Games as an indicia of white culture? The modern Olympics were born out of a European fascination with the Greco-Roman world, the same fascination that led to much of Western architecture, art, governmental systems, engineering, mathematics, science, philosophy, etc. One might argue that the Greeks, the Romans, and the

Europeans were all largely white people. Instead, I prefer to argue that they were all just people, humans, one species, thinking about their place in the world and leaving behind ideas and things like sports that are available to anyone willing to take up these ideas, work with them, practice, and engage. There is nothing about the 400-meter hurdles that benefits white people over black people. It is not racist. And a black man named Edwin Moses owned that race for a decade.

I don't mean to suggest that we don't have a problem, but the focus should be on fairness on who gets to run the race, enforcing the rules for competing so that no one has an unfair advantage and providing coaching for anyone who wants to run. Even if we do all of that, we will still have winners and losers because it's a race. That's the point. Not everyone is Edwin Moses. We can admire the winners, but we should love the losers too because they are human, and we all eventually lose.

In fields of human endeavor, like communicating with the written word or the scientific method, which the antiracists call part of white culture, the real issue is lack of access to education. None of us, black or white, are born knowing how to read, write, think analytically, or hurdle. We have to be taught those things, and we all have different levels of ability to excel at those things. We need a revolutionary recommitment to public education. We should pay teachers more and support early childhood education, especially in poor areas. Labeling those skills like writing and scientific reasoning as part of white culture diminishes them and those people, black and white, who have acquired those skills through hard work and education. If certain groups of people are disadvantaged because they don't excel at those skills, we should work at understanding why that is the case and attempt to remedy that problem, rather than concluding that writing and scientific reasoning are not important. They are important by their very nature, just as the 400-meter hurdles are important, because they unlock the infinite possibilities of human potential.

We owe the black community reparations for the horrific way that black Americans were treated in the past. By "we," I mean the United States of America. The United States has benefited so greatly from the slave labor in the early years of our country and from paying black people less than the value of their services after slavery ended, that in purely economic terms, it is unconscionable not to address that debt in some way. Everyone should listen to the 1619 podcast from *The New York Times*. It provides some history that makes this obligation clear. Of course, the political challenge to doing anything will be great. In my view, reparations should be paid in part with investment in public schools in predominantly black communities. The schools wouldn't be segregated; white students could attend. But they should be built where the black people are. We should create centers of learning that are so vibrant and progressive that the white folks farther away want to have their kids bused in. We should give every black kid who wants to go to college a scholarship to any public university in the country. We could take the money out of national security or defense budget because it will improve our national security.

Chapter 19:

Time

Theoretical physicists question whether time is anything more than just a construct that we humans have created. Some say that past, present, and future are one and that our perception of the flow of time is an illusion resulting from the limitations of our consciousness. As I understand it, heat is a critical factor because there is a detectable difference between past, present, and future only when heat exists. Thus, our illusion of time results from our limited consciousness swirling around in a soup of quantum variables that is changing only because of heat. We perceive these changes as the flow of time.

Occasionally, when traveling by train, I'll get the sensation that the train is moving only to find that it is the train next to us, visible through the window, that has begun moving. I imagine the argument about the flow of time is similar. We are stationary—i.e., there is no flow of time—but we perceive the flow of time because the world around us, including our own bodies, is changing.

Stephen Hawking made important contributions to theoretical physics while suffering from ALS and, remarkably, lived with the disease for over 50 years, dying at age 76 in March 2018. He observed that black holes emit heat like a stove, adding to our incomplete knowledge about time, space, and matter with reference to black holes. Our experience in this world is so dependent on our notions of past, present, and future that we can't discuss our world or orient ourselves within it without reference to time, as in the first sentence of this paragraph. I am encouraged that Mr. Hawking lived 50 years with ALS, but is the truth that he and I, and all of human kind, exist in the same instant? Do we only deceive ourselves with our notions of past, present, and future?

While theoretical physicists have the view that humans are made of the same quantum particles that constitute all matter, and behave subject to the same laws of nature that govern the rest of the universe, most recognize that our ability to be both part of nature and simultaneously observe and investigate nature is at least a source of wonder. Theoretical physics doesn't have much to say about our sense that we are part of nature but also somehow separate from nature, or about right and wrong, love and hate, art and music. For these things we turn to religion, literature, art, history, politics, and law.

Deepak Chopra, the well-known author and physician, wrestles with the intersection between science and religion in his recent book *Metahuman*. His struggle to reconcile the hard-won advancements in our understanding of the world through science with ancient wisdom handed down through religion and philosophy is admirable. He proposes that our human consciousness, our self-awareness and ability to observe the world around us, is part of a universal consciousness that is shared with all life. He points out that some animals seem to be aware of themselves as evidenced by how they interact with mirrors, and anyone who loves a pet can also attest to observing some level of consciousness in their pet. Our daughter has sent us video clips of her cat, Evie, interacting with a mirror in a way that suggested the cat understood the image in the mirror was herself. I believe we as a society will come to regret the way we have treated animals as we come to understand more about them and how much they share in common with us.

Dr. Chopra asserts that this consciousness, the one universal consciousness that we all share, is at the root of creation. As support for this, he goes back to quantum physics. According to quantum physics, particles are actually the collapse of energy waves. Standard quantum theory holds that particles exist in a virtual state, invisible and having no fixed location, until an observer causes the waves to collapse and particles to come into existence. Thus, at the quantum level, an observer, or consciousness, is necessary for the collapse of energy into matter, which sounds to me like creation.

I understand that this has been a controversial point in quantum physics, however, since the time of Einstein. Austrian physicist Erwin Schrodinger, a contemporary of Einstein, attempting to point out the absurdity of the theory as applied to everyday life, proposed a thought experiment that became famous as the paradox of Schrodinger's cat. Schrodinger proposed that if a cat were placed in a steel box with a gadget that released poison when a quantum particle came into existence by the decay of an atom, and if during a one-hour period the probability of that decay occurring was 50%, then during the hour the cat should be both dead and alive. Einstein agreed that there was a gap between the theory and real life.

Those of us who are not scientists are left to pick up bits here and there from books or the news and let it sink in, fill us with awe, and stretch our understanding of the world around us and our place in it.

Chapter 20:

Open Label Extension

My last drug administration LP in the trial was on December 20, 2018, and because we thought the drug was working, we were hoping for an open label extension (OLE) of the trial. OLE studies are often conducted after randomized, placebo-controlled drug trials to allow investigators to test the drug for safety and tolerability over a longer term than the placebo-controlled trial (two years is not uncommon, but they often continue until the drug is licensed). Participants in the placebo-controlled trial are typically invited to participate, and everyone is given the drug to evaluate its safety and tolerability in day-to-day use. Of course, the OLE must be approved by the FDA based on the information from the animal studies and human trial.

Laura heard through the grapevine that the OLE would not take place for several months after the end of the placebo-controlled trial because additional animal studies were needed for the FDA to approve longer term studies. Apparently, additional primate studies were needed, which would take several months. As a result, the OLE was not expected to take place until late 2019.

We were on our own again, at least for a time. I started taking the metformin again. On October 7, 2019, we went back to Baltimore to see Jeff Rothstein for a check-up. He did the push-pull neuro exam, and we asked about the OLE. He said he didn't know but thought it would likely be in the new year.

Before the OLE, I would have liked to have more information about the animal studies and the results of the placebo-controlled trial, but none of that was available. For example, why were additional animal studies required? Did the drug have some adverse effect? Was anything similar noted in the human trial? In the human trial, did the drug reduce the RAN proteins in the CSF or the blood? What indicia of effectiveness were noted? Were there any adverse effects? The FDA will have all that information, of course, and will use it to evaluate whether to allow the OLE, but it seems appropriate to provide it to the patients as well to support their decision to participate in the OLE. Based on this lack of disclosure, I could have declined to participate in the OLE and waited for more information to become available, but for ALS patients, time is precious. Based on my experience in the placebo-controlled trial, I decided I would participate in the OLE when and if it was offered.

We checked again in January 2020; there was still no word on when there might be an OLE.

Then COVID-19 arrived, and everything stopped while people complied with stay-at-home orders and figured out how to work in the new world. On June 29, 2020,

Alpa emailed me to ask if I wanted to have a video check-up appointment. I replied that although it would be lovely to see her, I didn't think it would be necessary, but also asked if she heard about the OLE. She referred me to Kristen Riley, who emailed that they were just waiting for the Institutional Review Board (IRB) to sign off on in-person visits for the study and for Biogen to officially open their site. The IRB committee at Johns Hopkins University needs to review, approve (or reject), and monitor all biomedical research involving human subjects. Its purpose is to assure that appropriate steps are taken to protect the rights and welfare of me and the other human subjects in the trial.

Institutional Review Boards were required by federal law in the wake of several notorious abuses of patient rights in clinical trials. After World War II, the United States tried many Nazi doctors for the barbaric and inhuman experimentation that had been conducted on Jewish people and other ethnic minorities in the German concentration camps. Also, from 1932 to 1972, doctors at the U.S. Public Health Service conducted the infamous Tuskegee Study of Untreated Syphilis in African American males. Black men in the study were told that they were receiving free health care from the U.S. government when in fact the purpose of the study was to observe the progress of untreated syphilis. The National Research Act was adopted in 1974, and subsequent regulatory measures required an Institutional Review Board at every institution that conducts research with human subjects.

We inquired about getting Vickie into the Biogen trial and were told that Biogen hasn't let Johns Hopkins or other centers participating in the trial know yet what their numbers of patients will be for the next cohort in the study.

As the summer of 2020 wore on in Florida, we had both been feeling a little trapped in our COVID confinement, even though we had every comfort, and even a swimming pool, at our disposal. Laura was more affected by this malaise than I was and started to investigate renting a recreational vehicle to drive from Florida to Minnesota to see Paul and Susan, and even Maddie, who by the end of summer had again left New York. Laura considered this the safest way to travel. We could pack all our food, meaning no stops at restaurants. We would have an onboard bathroom, meaning no stops for potty breaks. We would have a bed in the vehicle, meaning no stops at hotels. We would have to stop for gas, but we could wear gloves and a mask, and that seemed to be a manageable risk. All of this seemed preferable to air travel where we would be exposed to hundreds of people.

We rented an RV from Cruise America, which in Gainesville is represented by an undistinguished repair shop with a couple of guys working on cars. In the parking lot, however, were two RVs with photos of the magnificent red towers of Utah's Red Mesa emblazoned on the sides and back. We got the one that the tall, thin, young man with a black mask and big work boots told us was the better of the two, since, as he said, "you're planning to have it for quite a while." We rented it on August 27 and planned to return it on October 5.

Laura sanitized the vehicle, wiping every surface and control with a bleach solution, and then we plugged it into our house power with an extension cord because it takes overnight for the refrigerator to cool down. We had planned to leave the next morning, but we had badly underestimated the time necessary to load the RV with all our stuff and food for the trip. We finally departed at about 3:00 p.m. and made it to an RV park outside of Atlanta for the night.

It was a pretty good vehicle, as we learned on our trip across America, except for the damaged hose connector for the city water, and the lack of a shower curtain. At our RV park on the first night we spent about an hour trying to get the water connected and eventually just accepted that it was going to leak. The pressure in the RV was okay, however.

The second night we stopped near Chattanooga and visited the Chickamauga battlefield. I had been reading Ron Chernow's *Grant*, and it was moving to stand at the spot where the two armies, all Americans, had clashed and so many had died. I imagined boys as young as our big-booted Cruise America rep swept up and lost in that conflagration, and I imagined it happening as I stood there. Accepting for the moment that time is an illusion, I imagined platoons of ghosts charging, firing guns, and shouting, others dead or wounded and dying. Yet around me the wind whispered softly and dappled light shone on the hallowed ground.

We put in a long day of driving on the third day of our trip, stopping in Bloomington, Illinois, to have a socially distanced visit with Laura's brother, Bruce, and his wife, Dee. Dee has COPD, and because of her higher risk status with respect to COVID, they are especially careful. We all wore masks and face shields and chatted in the backyard from chairs about 15 feet apart. We found an RV park outside Bloomington for our third night.

We arrived at the Wisconsin lake home of Laura's brother, Clark, and his wife, Cheryl, on the afternoon of our fourth day. Clark and Cheryl have a new grandchild, Graham, who was born at 25 weeks in the fall of 2019, and although Graham is making good progress, he is still fragile. Cheryl is providing childcare so that Graham's parents, Charlie and Jess, can work during the day, and they are all cautious about protecting Graham from the virus. They had all conferred about whether or not they were comfortable with letting us into their "bubble," i.e. interacting with us without masks and without social distancing, and based upon how careful we have been over the last few months they agreed that we were in. It was surprisingly wonderful to greet them with hugs and to be able to hold and admire Graham close up. I didn't realize how starved we were for social contact.

Clark and Cheryl's lake home is a beautiful place overlooking Deer Lake outside of St. Croix Falls, Wisconsin, about an hour's drive from the Twin Cities. The living room has big ecclesiastical beams supporting the high ceiling with a fireplace at one end of the nave and a bank of windows looking out over the lake at the other. We stayed there about two weeks until finally Laura felt as though we were overstaying our welcome. Since I don't do guilt anymore, I would have stayed longer, as it was a little slice of heaven. But after that we moved to an Airbnb in Minneapolis for a week so that Laura could focus on her grant deadline with fewer worries.

We were staying at the Airbnb apartment when we heard news about the OLE for the ASO trial on the same day that Ruth Bader Ginsburg died, September 18, 2020. That morning Laura had been corresponding by email with Jeff Rothstein about a different topic and asked if he had any news about the OLE. He copied Kristen Riley on his response, and she emailed back, "We just received the activation email 4 minutes ago." The long wait was finally over. We were pleased with the prospect of getting back on the drug that both of us believe is the best hope for reversing or stopping the progression of my disease.

Laura and I were out walking later that day, feeling good about the positive news regarding the trial, when our Apple watches brought us the sad news of RBG's death. We were stunned. 2020 had served up yet another disaster. Although Justice Ginsburg was caricatured as an "extreme liberal" by those on the right of the political spectrum, she has earned a place in history as a warrior for equal rights for all people. I wonder how women, especially, can criticize her as too liberal given all she has done for women to be granted equal rights in the workplace. Those lawyers who argue that constitutional interpretation should be guided by the words of the founders and what they meant at the time they wrote those words are called "originalists." Justice Ginsburg forced originalists to see that even though the founders were not likely considering women as among the persons deserving the "equal protection of the laws," our societal norms of fairness and justice have moved to a place where that interpretation is appropriate and necessary.

The OLE was going to require a lot of trips to Johns Hopkins in Baltimore. Kristen sent us a schedule that had a screening visit on October 5, then three visits every two weeks to administer the drug during a "loading period," followed by a visit every month. Although Kristen scheduled us through May 7, 2021, the consent for the OLE said the trial would continue 22 months after the loading period, with continuing evaluation after that. My participation in the trial was to last two years and three months.

Laura called our Cruise America rental place and extended our reservation so that we could drive from Minneapolis to Baltimore for my October 5 appointment. Given that we didn't expect the COVID pandemic to be resolved anytime soon, we began to consider buying an RV to use for the upcoming trips from Gainesville to Baltimore. I spent a lot of time in that Airbnb learning about RVs and watching video reviews of RVs.

We spent our last day in Minnesota/Wisconsin with Clark and Cheryl at their lake place. The fall color was beginning to appear among the trees, and the air was crisp and cool, but warmed by a strong sun, all evoking past autumns with school, pumpkins, and football. Clark is still deep in the legal weeds, and Cheryl has a demanding schedule caring for Graham, and so they have their own battles to fight. Despite this, they have been generous to us, both with their hospitality and their concern about our welfare in this fight with ALS. We hugged them goodbye on the morning of Saturday, October 3, and started our drive to Baltimore.

Laura does most of the driving. For some reason, she feels more comfortable if I'm not driving. I've asked her what she's concerned about, and she talks about my reactions when startled or my sneezes. I do have a stronger startle response than I had before ALS. If I am sitting at my desk sipping hot coffee focused on something and the phone rings, I jerk so violently that I spill the coffee.

My sneezes also seem to be more violent. One day I was sitting in my office chair looking at my cell phone when I sneezed so violently that I fell forward out of the chair and bruised my head on the floor. I was unable to break my fall with my hands because I hesitated to drop the phone. If you ever find yourself choosing to save either your phone or your head, I advise you to choose the head.

Although I think I can drive perfectly well, she may have a point. I am perfectly happy not driving also, since it allows me to read or nap, unless Laura has me navigating or tuning the radio or finding a cord to charge her phone or playing a podcast, or finding a place to stop for the night, etc. She has a lot of demands.

The screening visit on October 5 was uneventful. They took my vitals and blood and urine samples, then we walked out of Johns Hopkins, climbed into our Cruise America RV, and headed home. Not long after we returned to Gainesville, we decided we needed to buy an RV for the many trips to Baltimore ahead. We assumed we would not be flying for some time. Before our next appointment on October 19, a 24-foot, Class B RV, a Coachmen Galleria, was parked in our driveway, and we were ready to begin our experience as road warriors. We drove round-trip from Gainesville to Baltimore, about 1,700 miles, eight times from October 2020 to April 2021. We listened to some good audiobooks, cringed at the Confederate flags along I-95, and got to know which RV parks were better than others. It's a long trip, and it takes at least four days for the round trip.

We were in the van, driving to Baltimore on March 11, 2021, when we got a text from Vickie. She had sent a link to an interview titled "How COVID-19 'Vaccines' May Destroy the Lives of Millions" by Dr. Joseph Mercola. In the text, Vickie said, "I would love your honest opinion on this. Does this make any sense?"

The link brought up a brief introduction by Dr. Mercola in which he says animatedly that "we are seeing people die as a result of this intervention, and we are going to see many, many more deaths and even worse, permanent disabilities" and announces his interview with "one of the leading experts in the entire world," Dr. Judy Mikovits. Dr. Mikovits appears in a baseball cap and says, "I'm just actually literally beside myself with anger over this gene therapy, this synthetic chemical poison, and what they're doing worldwide. We are already seeing the victims, the deaths from this shot. It's illegal. It shouldn't be done. It should be stopped right now. It should never have been allowed to happen." She goes on to say that the vaccine will lead to more neurological disease like Parkinson's and, significantly, ALS.

We had earlier encouraged Vickie to get the vaccine so that she could feel more comfortable traveling to Johns Hopkins to see Dr. Rothstein with the hope that Dr. Rothstein could get her into the clinical trial that I was in or a clinical trial for another ALS drug. After Florida opened up vaccines for people vulnerable to COVID on March 1, I got my general practitioner to sign off that I was vulnerable and got my first dose of the Moderna vaccine on March 5. We were all in on the vaccine and were surprised that Vickie was hesitant to get the vaccine.

But this is a good example of how different our worlds are. When I look at that video, my skeptical radar is immediately engaged by the lack of academic seriousness of the discussion. Dr. Mercola looks like he is in an ad for a horror movie, with wide eyes while he predicts many deaths and disabilities. Dr. Mikovits doesn't look like a real scientist, wearing a baseball cap and making statements about poison and deaths. There is no nuance, no recognition that the truth is rarely black or white, but rather shades of gray. Then I Googled Dr. Mercola and Dr. Mikovits and saw that they are somewhat controversial and not in the mainstream of medical opinion. As a result, I didn't believe what they had to say and assume they were seeking publicity to sell products or books or both.

But Vickie didn't see that. I texted back, "No, this makes no sense. The vaccines have been tested on tens of thousands and now given to millions. There have been no reports of destroyed lives." But she never got the vaccine. I assume she was listening to others with different opinions.

Maybe she was talking to Cal. On March 17, I received a text from Cal with a link to another anti-vax video. She texted, "Hey dear brother, please listen to this and tell me what you think." In the video, Dr. Sherri Tenpenny says, "Some people are going to die from the vaccine directly. But a large number of people are going to start getting horribly sick and get all kinds of autoimmune diseases." I Googled Dr. Sherri Tenpenny and quickly found an article on politifact.com that rated the claims in the video "Pants on Fire" false, and sent it to Cal.

I will respond to my sisters' texts and offer my opinion when asked, but I don't expect it will do any good. I got my second dose of the vaccine on April 5, 2021 and a booster on October 24, 2021.

We started flying to our appointments on May 7. Laura and I agreed that flying to Baltimore was preferable to the two-day drive in the RV, and that the Inn at Henderson's wharf was preferable to RV camping in KOA's off the interstate. Wearing masks for hours on the airplane and in the airport was unpleasant but an acceptable price to pay for the benefit of flying. Our time in the Atlanta airport was improved by visiting the Delta lounge on the F concourse which has a sort of open-air patio where we could take our masks off and enjoy some snacks and something to drink. We stopped having our groceries delivered and no longer wipe them down before putting them away into the house. It was actually fun to go back to the grocery store and pick out the items ourselves, even though we were wearing our masks. We began to feel as though the pandemic was slowly ending.

Chapter 21:

Fourth Of July, 2021

For the Fourth of July, we decided to travel from Baltimore, where we had an appointment at Johns Hopkins on July 2nd, to Minnesota to see our kids and to visit Vickie whose condition was becoming increasingly worrisome. Maddie had resigned from the horrible Manhattan law job she had started in October of last year and flew to Minnesota also to meet us. Laura and I met Maddie at the Minneapolis airport, rented a Toyota Rav 4 and drove to Paul and Susan's house in St. Paul.

We spent the night there and on the morning of July 3rd, the five of us, (me, Laura, Maddie, Paul and Susan) were planning to drive to Bridget and Ryan's house in Cold Spring, Minnesota where Vickie and her husband Steve were expected that day. Vickie's other daughters Becky and Sarah were also planning to be there with their husbands and kids. All 11 of Vickie's grandkids would be there. Bridget and Ryan live on a lake and so there would be swimming and watersports to entertain the kids.

Paul and Susan made waffles, scrambled eggs, bacon, fruit salad and coffee for breakfast that morning and spread the serving dishes out on the kitchen counter for a buffet service. Their kitchen has lots of south facing windows and so was sunny and cheerful. We were chatting and serving ourselves, and I was about to walk out to the deck to sit at the table and chairs there when Paul picked up a piece of paper from among the serving dishes and said, "Guys, um, take a look at this photo." I was standing a bit to the side and didn't see what he was holding, but I saw the shocked reaction of Laura and Maddie. Laura grabbed the paper with some urgency as she took a sharp, audible intake of breath and then began to laugh or cry or maybe both at the same time. Maddie simultaneously gasped, and said slowly and with a high-pitched voice, "Suuusaaan! Oh, my God!" It was an ultrasound image of a fetus. Susan was pregnant! Everyone started to laugh and hug. Maddie held her face with both hands and said "Holy shit!" As I hugged Susan, I said "Wow, Susan, I am so happy!" I hugged Paul tightly and said "Congratulations! Wonderful!" I was thrilled. We were going to be Grandparents. A new life was starting.

Susan said, laughing, "We tried to slip it in but no one saw it. It was just sitting there." Laura and Maddie were studying the image. Susan asked, "Can you make it out?"

"So, its little face is there? Maddie said, "and then its little foot? Oh, my God. Wait, so how many weeks slash months?"

"Twelve weeks" Susan said, "So that was just last week."

"So you waited three months?" Maddie asked.

"We wanted to tell you guys in person."

There was a pause while people wiped tears and gathered themselves. Susan said "So, January, 2022"

Paul added, "Coming in hot." We all laughed more.

Eventually, Laura asked "Is it a boy or a girl, or do you know or you don't want to say?"

"Yes, we know" Paul and Susan said in unison.

Maddie involuntarily tensed her arms in front of her, raised one knee and said "ooohhh" then held her face with both hands, and added "okay, wait, let's take bets."

Laura, the scientist, studied the ultrasound image looking for data, raised her glasses and brought it closer to her eyes. She said, "well, I just don't know."

Paul said, "I don't think you're going to be able to tell anything from that photo."

"I'm guessing a boy." Maddie said assertively.

"Feel good about that?" Susan asked mischievously.

"I mean I don't feel great about it now that you've said that, but …." Maddie responded.

Laura interrupted her, "Well, I'll guess a girl, then" as she put an arm around Maddie.

I said, "The Ranum family leans girls," thinking of my brother's and sisters' kids which consist of eight girls and no boys.

"Is that your guess?" Susan asked.

"Yes, I guess girl." I said.

"So, girl, girl, boy" Susan said, pointing to me, Laura and Maddie, respectively.

Susan looked at Paul, letting the tension build. Laura, Maddie and I leaned forward expectantly.

"It's a boy!" they said in unison.

We threw our arms up in celebration and repeated "It's a boy!" giddy with laughter at the wonderful news.

Maddie said with conviction, "I knew it!"

Some news arrives with such force that it splits time into before and after. This was that kind of news. Now that we were aware of that little boy growing safe inside Susan's belly, things were different. He occupied our minds tenaciously. In the days that followed Laura and I were constantly asking each other "Can you believe we're having a grandbaby?"

We left after the announcement and breakfast for Bridget's house about an hour and half's drive away. We took two cars so that Susan could travel back separately if she were called back to work. Upon arriving we hugged everyone, and as I hugged Vickie I was startled by how thin she was. I could feel the bones beneath her skin. She later told me her weight was down to 91 pounds. Vickie is about five foot three inches and I understand before she started losing weight from ALS she was about 137 pounds.

Despite being so thin, she looked good. She wore a new blue dress for the occasion, with red and white panels on the front in the spirit of the holiday. She has always been pretty and one of the things that I've heard her complain about is the difficulty she has dealing with her hair now that her arms are weak and difficult to raise to her head. That day she wore makeup and lipstick and her blond hair was neat with wispy bangs and locks that fell a few inches past her shoulders.

She sat on the couch with a kitten she had named Buttercup. Buttercup had gotten a thorn in one of her front legs and was avoiding putting any weight on that leg. I sat beside Vickie and Susan and Maddie sat on the other side of her. We were all aware that this might be the last time we saw Vickie. We chatted about the kitten and the amazing spread of food that Vickie's daughters had laid out on the table. There was a separate table for desert featuring a strawberry/rhubarb pie and adorned with store-bought wrapped treats of various kinds.

"How was your drive down?" I asked Vickie.

"Oh, it was long." Vickie said.

"Were you able to sleep in the car?"

"No, I couldn't get comfortable sitting in that car. I don't think we'll drive down here again." There was a pause during which she apparently concluded that sounded unacceptably hopeless because she added, "this year."

Attempting to turn the conversation in a positive direction, I said, "I'm sorry it was a hard trip, but it must be great to see your daughters and all these grandchildren. What a beautiful family you have!"

She smiled and said, "Yes." Then her expression suddenly became more serious as she turned to Susan and said, "when are you and Paul going to have a baby?"

Susan laughed and said "Well, Paul and I do have some news to share. We're expecting! The baby is due in January."

Having gotten the answer she wanted, Vickie was all smiles as she hugged Susan and said "That's great! What happy news!" Becky and Bridget who were nearby in the kitchen apparently overheard this and appeared to add their congratulations and the news quickly circulated throughout the house.

Even if Vickie and I die from the ALS that we struggle with, we will continue through our children and grandchildren. A piece of us will go on. This is the nature of life. And the part that causes ALS, the mutated C9orf72 gene, will not go on. It will die with Vickie and me. That is a great comfort to me and I'm sure it is to Vickie as well. We all want life to flourish without suffering or difficulty.

After lunch, that instinct drove Susan and Maddie to do something about Buttercup. Vickie said they had tried to pull the thorn out but a piece must have been left behind because the right front leg was clearly infected. Susan examined the kitten and concluded that the abscess that had developed around the wound needed to be lanced and the pus drained.

After a quick trip to local drug store for supplies, Susan and Maddie set to work. As Maddie held the kitten, Susan shaved an area around the wound and wiped it with alcohol. The kitten tolerated this without much protest. Then Susan cut a small incision in the abscess with a razor blade and the kitten mewed with complaint despite Maddie's best efforts to comfort her. The mewing continued with greater urgency as Susan squeezed a significant amount of pus from the incision, wiping it up with paper towels. When they were done, Buttercup was relieved to be released from Maddie's grip. Susan was unsure if the wound would heal and advised Vickie to take Buttercup to a vet when they got home. We later received a report that Buttercup had fully recovered without a visit to the vet thanks to Susan's good work.

Later we took a lovely pontoon ride on the lake which was busy with others enjoying the weekend holiday. We admired Ryan's second "pontoon", which consisted of eight beer kegs—four on each side--strapped to a picnic table and powered by a trolling motor. After the pontoon ride, Laura, Paul, Susan, Maddie, and I said our goodbyes, did another round of hugging, and started the approximately two-hour drive to Mike and Mary's lake home near Park Rapids.

When we arrived at Mike and Mary's place, the celebrity of the visit, twelve-month-old Ella, was already asleep. Her parents, Emma and Matt Erickson, had brought Ella for a holiday visit with grandparents Mike and Mary at their magnificent, timber-framed, lake home. After a wonderful dinner and a good night's sleep, we woke to the sound of Ella's running footsteps on the floor above us and I thought *pitter patter of little feet is a pretty good description.*

Ella is absolutely adorable and Laura and I see a little of grandmother Mary in that young face.

Later in the week, the five of us visited Clark and Cheryl at their lake home in Wisconsin, completing our trifecta of Minnesota/Wisconsin lake visits on this trip. There we saw Embry, whose parents are Clark and Cheryl; Page and Stephen, Laura and Clark's niece and her husband: Eva and Truman, Stephen's children and Page's stepchildren; Charlie and Jess, Clark and Cheryl's son and his wife; Graham, Charlie and Jess's son; and Ray and Maureen, Cheryl's mother and father.

Since Page is also pregnant and farther along than Susan, they seemed to enjoy comparing their bumps.

This is life. It involves lots of people, fetuses still just a bump in their mama's belly, pre-teen kids finding out who they are, young adults finding their way in the world, parents balancing careers and families, grandparents, etc. It involves beautiful lakes, boats, waterskiing and fourth of July gatherings. And at some point, it involves sickness and death. It is all good, wonderful and miraculous. I see that more clearly now than when I was engaged in the grind of the legal practice and worried about getting clients and making money. While it is tempting to embrace self pity, that seems petty and self-absorbed when I consider the wonder of creation and the fact that all living things die. Having been part of this miracle of life and so richly blessed in this world, I expect that my passing from this world will be equally miraculous and wonderful.

For those of us in the countries with good medical care, we don't have to fear much pain. They have means to control that. But good medical care can also prolong life long past the point where the reason for living has been left behind. This is the real issue with ALS. Patients with ALS often get a feeding tube and then a ventilator and continue to live unable to move unassisted and hooked up to machines. Is that a desirable alternative to death? That is the difficult question that Vickie would have to confront sooner than we realized and that I will have to confront eventually.

Our mother did not confront that question because she did not have good medical care in 1989; no one had advised her or my family that she was at risk of dying while she slept. ALS often affects the diaphragm and muscles involved in breathing, and I assume mom simply stopped breathing. I don't know if there was a moment of panic or if she passed peacefully moving toward a bright light emanating a feeling of love, compassion and unity. But her death seems a good death except for the fact that it was unexpected.

Still, even assuming we hang on a bit too long, and our final days leave little to be enjoyed, the end is not likely to be horrific. By comparison with the Serengeti, which we visited in 2020 before COVID, and the lives of the gazelles or the warthogs that were ended in gruesome fashion by lions, we have little to fear.

Chapter 22:

Vickie's Final Days

On Monday evening, August 2, 2021, I got an alarming text from Vickie: "I could use a little help. I have this phlegm; I'm trying to cough it up, but it's not coming." I called her right away.

"Hi, Vickie, I saw your text. Are you able to talk?" I asked.

"Yes, I can talk." Her voice was weak and she continued haltingly, "I ate a couple of Brazil nuts, and now I have this phlegm in my throat that I can't get up. Steve made me some tea with ginger and now I'm sipping that. My cough is so weak."

I was relieved that she could talk. "Good, sipping something should help." I said in my slow, ALS-affected voice. "I've had a catch in my throat that causes me to cough after eating nuts also and drinking something usually helps." I thought, *what a pair we are with our feeble voices, comparing our problems with eating nuts.* Laura was with me and I had the phone on the speaker. She told a story about my coughing at the same cloverleaf on the highway in Baltimore on separate trips to Johns Hopkins because each time we left the appointment hungry and had snacked on nuts as we drove out of town.

"Do you still have an urge to cough?" I asked.

"A little." She responded.

We told her to keep sipping something and said we hoped that she felt better soon. We asked her to keep us posted as to how she was doing. We received additional text reports in which she reported her condition as "Fair" and said that although she still had an urge to cough, it was not as bad. We went to bed thinking it would probably resolve overnight.

When I called in the morning, I learned it had been a rough night for Vickie. She experienced coughing episodes lying flat in her bed and so had spent most of the night dozing in a chair. She also said that her finger pulse oximeter had reported a blood oxygen level of 77% and that she had a pain in her side. Having been sensitized to the dangers of low blood oxygen from news about COVID, I immediately said to Vickie that was too low and she should call her doctor. She had some kind of oxygen machine at home and later, after napping with that machine, she reported that her blood oxygen level was back up to 85%. I didn't get a response to my text in the afternoon asking if she had reached her doctor.

The next day, Wednesday, Vickie and Steve got in to see the doctor, and Vickie had a chest xray that disclosed that she had pneumonia. Given her ALS and the pneumonia, the doctor prescribed antibiotics and advised Vickie to go to the hospital where they could monitor her more closely. Vickie was concerned about "all the germs" in a hospital setting and said she would prefer to go home where she expected the antibiotics would improve her condition soon. Bridget and her daughter, Scarlett, arrived Wednesday evening to help out. We were pleased that Bridget was with Vickie to comfort and support her and spoke with Bridget by phone that evening. Vickie was there also but had the oxygen machine mask on and did not talk. It seemed as though things were in a good place for the evening, and we all hoped the antibiotics would kick in soon.

On Thursday, things took a turn for the worse. Bridget and Scarlett were with Vickie and together they called the doctor when Vickie didn't seem to be improving. The doctor urged them to go to the hospital. The emergency room put Vickie on a BiPAP machine and did a routine COVID test. BiPAP stands for bilevel positive airway pressure and the machine increases the pressure when the patient inhales and reduces the pressure when the patient exhales while also enriching the air with oxygen, thereby making it easier to breathe. It's noninvasive and delivers the air through a mask. Surprisingly, Vickie tested positive for COVID. Then they tested Steve and he was also positive. Vickie now had to fight both ALS and COVID.

The hospital moved her to the COVID ward in the intensive care unit (ICU) and tragically, Vickie lost the thing she valued most, the comfort of having her family by her side. Family was unable to visit.

On Thursday evening at 9:41 pm, I got a text from Vickie, "If they need to incubate *[sic]* me should I. Let *[sic]* them." I was tempted to say "No!" but then I thought of Steve Gleason and his "No White Flags" motto and Stephen Hawking and his productivity long after getting on a ventilator. *Is refusing intubation a surrender?* I thought also of Atul Gawande's wonderful book "Being Mortal" which asks whether the quality of life after expensive and invasive life-saving measures justifies all this medical intervention. Susan says that many of the patients who are intubated in the ICU just stay there longer and die more slowly. Vickie was so weak that I doubted that she would recover enough to get extubated and breathe without assistance.

But regardless of these concerns, I knew that Vickie had to make this decision on her own. I responded with the following text, "You have to make that decision, Vickie. I am not sure what I would do in your place. Have you talked to your doctor about that?" Vickie did not respond. That was the last text I received from her.

At about midnight on Thursday, Becky texted that the ICU doctor had confirmed that Vickie's pneumonia was not due to COVID and was most likely due to food aspiration. We were encouraged by this and hoped the antibiotic might help Vickie turn the corner. Becky also reported that the ICU doctor had asked Vickie and the family to think about whether Vickie wanted to take more invasive measures if needed like intubation and resuscitation if her heart stopped, which I realized must have been the conversation that prompted Vickie to send the text earlier.

Since my news about Vickie was a text message stream from Becky, Bridget, and Sarah and their access was limited after Vickie was moved to the COVID ward, the news slowed down significantly. On Friday about noon Becky texted that Vickie had changed her code status to DNR/DNI or do not resuscitate/do not intubate. Becky reported that Vickie had had a long discussion with the ICU staff. I was proud of Vickie for making that decision and thankful to those good doctors and nurses on the ICU staff for helping her. Medicine requires compassion and effective communication skills as well as the science.

The news was up and down. On Saturday about noon, Becky sent the following text, "I just spoke with her. She is miserable. Horrible headache, arms and hands weaker, couldn't reach the buzzer, phone or mouth swab. She says thank you for your prayers and love to you all." We were all discouraged. Bridget reported at about 9:00 pm that Sarah had talked to the ICU nurse who said that after testing Vickie's swallowing, they gave her some food and water and that went well. Her blood oxygen saturation was holding at about 98% and her blood pressure was stable. The nurse also said that Vickie seemed to be doing better vs. worse. We were all encouraged.

Sunday morning Vickie had coffee, oatmeal, and part of a muffin according to a "really sweet nurse" who had talked to Sarah. The nurse also reported that Vickie seemed irritated and had not slept well. A physical therapist was coming to work with Vickie on some exercises to enhance movement. That breakfast, just by the ordinary sound of it, and the idea of Vickie working with a physical therapist made me think *maybe Vickie will bounce back and be able to go home*. Later in the day, Bridget FaceTimed Vickie and sent a photo of her in her hospital bed with a breathing tube in her nose and looking tired and unhappy.

Over the next week, Vickie's lungs and kidneys developed problems, and although the doctors did not attribute these problems to COVID, we wondered what role the virus was playing. COVID is known to attack both the lungs and kidneys. She had a bronchoscopy to remove a mucus plug from one of her lungs. She was isolated in the COVID ward and unable to see family.

On Friday, the family was allowed to visit Vickie. We received a photo via text of Bridget and Sarah sitting at Vickie's bedside with their masks down below their chins. We worried that Vickie might still be capable of transmitting COVID but assumed the hospital staff must have tested her to make sure she was not contagious.

The hospital only allowed two visitors in the room, so some combination of Becky, Bridget, or Sarah were with Vickie all day. I'm sure it gave Vickie great comfort to have her daughters with her again.

But her kidneys remained a concern. The doctors planned to put in a line to start dialysis if they did not show improvement. Vickie's heart rate and blood pressure were high. Becky wrote in her text, "Can barely open her eyes. Eating and drinking very little. She keeps refusing the in-the-nose feeding tube. She said she wants to go home, but also says she wants to live. Thank you for your continued prayers."

On Saturday, Vickie consented to the in-the-nose feeding tube, also referred to as a nasogastric or ng feeding tube. She needed nutrition, and the doctors had determined she was unable to swallow safely. The kidneys had not gotten any worse, which the family took as a good sign.

Bridget texted on Sunday about noon, "Mom has more life to her today! She verbalized she feels stronger. And we even got a few smiles. It was very encouraging to see. She is listening to church and feels relaxed. Her BP is really good right now. Her favorite hymn is Blessed Assurance by the way."

Bridget also offered a FaceTime with Vickie, and we responded that we would like to do that. When they called us for the FaceTime, Bridget and Sarah were with

Vickie and held the phone up so that we could see Vickie's face. It was good see her alert and communicative.

Laura and I told her that we loved her, that we were so glad that she was feeling stronger and that we hoped she could go home soon. Vickie said in a weak voice that she loved us too. We didn't want to keep her on the phone too long and we were saying our goodbyes when Vickie said, "oh wait, did you know that Emma's expecting again?" We laughed and said that was news to us. We were encouraged by the call and reassured by Vickie's unflagging interest in the extended family developments. We thought maybe she would beat this crisis and have more good months at home.

On Tuesday, Steve and Vickie were able to see each other for the first time since they had tested positive for COVID. Becky reported that Vickie was not getting better, "The doctors think she may have a week if we bring her home. The hospital is keeping her alive. She is getting weaker and needs more help with her breathing. Her best day was Sunday." Still optimistic after Sunday's call, we were surprised by the foreboding tone of the message.

On Wednesday, Becky texted that they planned to bring Vickie home on Thursday. In the evening she reported that Vickie had been sleeping the entire day and had not awakened once.

On Thursday, August 19, my phone rang and when I saw that Rebecca Seeger was calling, my heart sank and I thought *oh no, she's gone.* Becky said in a teary voice, "Rob, your sister is with your mom and dad now." I knew that Becky's grief would be deeper than my own, and so I tried in vain to offer words of comfort. It was a short call because Becky had others to make. They were still at the hospital when Becky called. Someone had written on the white board in Vickie's room "Going home today."

Cal was traveling to Minnesota to be with Vickie when she received the news of Vickie's death and she continued on to help Vickie's daughters with funeral arrangements. The funeral was held on the following Tuesday. Laura and I flew from Gainesville to Minneapolis and Maddie flew from New York to Minneapolis on Monday. We all stayed with Paul and Susan in St. Paul Monday night.

In the morning Laura, Paul, Maddie, and I drove to Moorhead where we met Mike and Mary, Ruth and Nick and their daughter Elin, and Emma and Matt and their daughter Ella, for a picnic on the Concordia College campus, where Mike and Mary had met. Laura tested everyone of the 12 people at the picnic for COVID with a rapid 15 minute test so that we could all be comfortable not wearing masks. Our two families were more concerned about COVID than the rest of the family.

We did not attend the indoor service at Vickie's church but brought the live stream of the service up on one of the cell phones in the car as we drove from Concordia to Climax, Minnesota. It appeared to be a good crowd, but few, if any, were wearing masks. With the prevalence of the Delta variant of the virus and its rapid transmissibility, we concluded we had made the right choice to avoid the indoor service. We were planning, however, to attend the outdoor graveside service which was to begin at 4:30.

Vickie was buried in Birgit Cemetery near Climax, Minnesota, about a half mile from Vickie and Steve's house. It is a small cemetery carved out of the corner of a field beside a gravel road. The headstones largely have Scandinavian names like Erikson and Anderson although I did see a headstone with the name Hong. I wore a white shirt and tie with my jeans and suitcoat as well as a hat to shield me from the sun. It was hot, at least 90 degrees, and I soon regretted the coat but thought it proper to leave the coat on until the service was over. Our contingent stood out from the other guests as they arrived as we were wearing masks and the others were not. We avoided hugging and tried to chat from a distance with the relatives that we knew.

Mike and I had prepared some remarks to honor our sister, and Mike delivered them at the graveside. Vickie's daughters each said something also. A few others shared a few words. A consistent theme in everyone's comments was Vickie's kindness. Her extraordinary kindness left an impression on everyone she met.

In preparation for the funeral, Becky put together a memorial honoring those in our family who had died from ALS. It read, "We honor the memory of Vickie and those members of her family who also fought courageous battles with Amyotrophic Lateral Sclerosis, Lou Gehrig's Disease.

Muriel (Hoaas) Ranum
Vickie (Ranum) Stortroen
Lester Hoaas
Mildred (Hoaas) Haugen
Margorie (Haugen) Zahl
Deborah (Haugen) Gudmundson
Gladys (Hoaas) Haugen
Robin (Zahl) Ekren

Knowing that I will likely join that list one day, the eight names on that list have a hold on me. They are my mother, sister, aunts, uncle and cousins. Like them, I have the expansion mutation in the *C9orf72* gene. But I have a better chance to live longer than they did because of the clinical trial I'm participating in, the metformin I'm taking, and the lifestyle choices regarding diet and exercise that I've adopted. I hope it is many years before I join that exclusive club.

It is hard to believe Vickie's gone from this world. And who knows? Maybe she's not. There is a honeysuckle bush near the western end of pool, and sometimes as I reach that end of the pool and stop to catch my breath, I'll catch sight of a hummingbird flitting among the blossoms, brightly illuminated by the morning sun. *Maybe that's Vickie,* I've thought, *come back in another form, but still seeking the sweetness this world has to offer.*

Chapter 23:

Resolution

If you have come this far with me, dear reader, I owe you a resolution of my story. I will, like all of us, eventually die, whether from ALS or some other cause, but that is an unsatisfactory resolution because it seems, thankfully, to be more remote than I thought when I was first diagnosed five years ago. Then, I read the oft-repeated estimate that ALS patients die within three to five years of diagnosis and began to prepare myself for what I thought was a relatively short remaining time here on this earth. Now, five years from diagnosis as I write this, I have more hope for years of good life ahead because of the clinical trial with the ASO, the metformin that I take, and our efforts to eat and live in a way that is mindful of our health.

I would like to write a resolution to this memoir reporting that the ASO turns out to be a cure, but that seems to be a little premature. The clinical trial is still in Phase I, and Biogen will have to conduct a much larger Phase II trial and then a pivotal Phase III trial before the drug is approved by the FDA. There is a long way to go, and it may take several years before the drug is proven to be safe and effective. There is also the possibility that the ASO for *C9* ALS won't make it through the clinical trials and that Biogen will abandon the drug. Roche recently halted a Phase III trial for another ASO to treat Huntington's disease, demonstrating that similar drugs can go a long way down the clinical testing pipeline and still fail. Nevertheless, I am optimistic that the ASO for *C9* ALS will prove to be safe and effective, but that conclusion is too far off to allow me to make that the tidy end of this story. Maybe I'll be able to write a sequel to report the results.

The resolution I offer instead of death or a cure, dear reader, is a new and improved me. When we first moved to Florida, I worked hard to be a successful lawyer and spent a fair amount of time worrying about billable hours, client relationships, and income. A significant part of my ego was based upon being a good and successful lawyer. ALS forced me to reevaluate that, and it crumbled quickly under examination. It was a world of minutiae, and while it allowed me to make a fair trade of my effort and skill for economic advantage, the work itself was not particularly meaningful. What is meaningful are the relationships that I had and still have. If you think back over your life, which is more important: the accomplishments on your resume or the people whose lives intersected with yours? For me, Laura, Paul and Susan (and that as-yet-unnamed grandson in Susan's belly), and Maddie are everything.

This focus on what's important and on my place in this world has given me a greater appreciation and respect for all humans, animals, plants, and other life on this earth. What a miraculous, vibrant, pulsating mess we are! I have less fear of death. I am swept along in the tide of life. Despite all my efforts to stay healthy and enjoy my time here with my family, I know I really don't have any control, and my life will end sooner or later. The fact of that inevitability reduces the stress and fear of my circumstance. Although I will do my best to survive, I am more curious than worried about the end.

Sometimes, if I've been swimming regularly and am relaxed, I'll wake in the morning slowly and rest there in that delicious space between wakefulness and sleep, enjoying a meditative state without thought. Often, if I catch myself there, I'll try to imagine myself separating from my body and floating to the ceiling looking down at the bed with me in it, or even above the house, looking down on the roof and the pool in the back. I say "try" because I've never had an out-of-body experience, but I think of it as a good exercise to let go of the worries about that body of mine. I am not my body. *Or am I?* I am not so arrogant as to assert that I know the answer, but I want to believe that I am not my body.

Writing this memoir is another attempt to capture a piece of me that is not my body and send it off into the world. Sadly, the written word is to truth as a toy horse is to a real horse, but it is all we have. I am hoping to leave something behind that will remind my family how much I love them and tell the rest of those who might happen upon this book a little about my life. Not because my life was particularly important to anyone but me and my family, but because we all want to reach out to each other. I believe that's a good instinct. We should all reach out and connect and embrace our common humanity. There is much more that connects us than divides us.